Reference Card for Calculus I

This side of the reference card contains all of the major skills of Calculus I, as they are covered in this book. It provides you with an overview of everything you should know at a glance. Use it to make sure you have manitions, theorems, and techniques in every chapter in the first four chapters, while important, are not are omitted from this list.

Chapter 5
- ❏ Describe what is meant by a limit.
- ❏ Define right- and left-hand limits.
- ❏ When does a limit exist?
- ❏ Under what three conditions doesn't a limit exist?

Chapter 6
- ❏ Use the substitution method to calculate limits.
- ❏ Find limits by factoring.
- ❏ Find limits using the conjugate method.
- ❏ Use limits to find equations of vertical and horizontal asymptotes.
- ❏ Calculate limits at infinity.
- ❏ Identify and apply four special limit theorems.

Chapter 7
- ❏ Define a continuous function.
- ❏ Describe three types of discontinuity.
- ❏ Classify discontinuity as removable or nonremovable.
- ❏ Apply the Intermediate Value Theorem.

Chapter 8
- ❏ Write both versions of the difference quotient.
- ❏ Calculate general and/or specific derivative values using both difference quotients.

Chapter 9
- ❏ Explain where derivatives can't exist.
- ❏ Apply the Power Rule.
- ❏ Apply the Product Rule.
- ❏ Apply the Quotient Rule.
- ❏ Apply the Chain Rule.
- ❏ Find average and instantaneous rates of change.
- ❏ Give the derivatives of all six trigonometric functions.

Chapter 10
- ❏ Find equations of tangent and normal lines.
- ❏ Derive an equation implicitly.
- ❏ Differentiate an inverse function without first finding the inverse function.
- ❏ Derive a set of parametric equations.

Chapter 11
- ❏ Find relative and absolute extreme points.
- ❏ Find critical points for a function.
- ❏ Construct a wiggle graph.
- ❏ Apply the Extreme Value Theorem.
- ❏ Determine the direction and concavity of a function using its derivatives.
- ❏ Apply the Second Derivative Test to classify extrema.

Chapter 12
- ❏ Find the equation for velocity given position.
- ❏ Find acceleration given velocity.
- ❏ Write a position equation describing projectile motion.

Chapter 13
- ❏ Use L'Hôpital's Rule to evaluate limits.
- ❏ Apply the Mean Value Theorem.
- ❏ Apply Rolle's Theorem and explain how it differs from the Mean Value Theorem.
- ❏ Calculate related rates.
- ❏ Optimize a function.

ALPHA

Reference Card for Calculus II

This side of the reference card contains all of the major skills of Calculus II, as they are covered in this book. It provides you with an overview of everything you should know at a glance. Use it to make sure you have mastered all of the important definitions, theorems, and techniques in every chapter.

Chapter 14
- ❏ Calculate right, left, and midpoint Riemann sums.
- ❏ Approximate area using the Trapezoidal Rule.
- ❏ Approximate area using Simpson's Rule.

Chapter 15
- ❏ Apply the Power Rule for Integration.
- ❏ List the antiderivatives of the six major trigonometric functions.
- ❏ Apply the Fundamental Theorem of Calculus to calculate area.
- ❏ Use the Fundamental Theorem of Calculus to derive definite integrals containing variable limits of integration.
- ❏ Use u-substitution to integrate.

Chapter 16
- ❏ Find the area between two curves.
- ❏ Apply the Mean Value Theorem for Integration.
- ❏ Find the average value of a function.
- ❏ Calculate the distance traveled by an object.
- ❏ Evaluate and derive accumulation functions.

Chapter 17
- ❏ Integrate a fraction by separating it into smaller fractions.
- ❏ Rewrite an integral by first long dividing it.
- ❏ Utilize inverse trigonometric functions to integrate.
- ❏ Integrate by completing the square.

Chapter 18
- ❏ Integrate by parts via the brute force method.
- ❏ Integrate by parts via the tabular method.
- ❏ Integrate by first decomposing into partial fractions.
- ❏ Calculate the value of an improper integral.

Chapter 19
- ❏ Use the disk method to compute rotational volume.
- ❏ Use the washer method to compute rotational volume.
- ❏ Use the shell method to compute rotational volume.
- ❏ Calculate rectangular arc length.
- ❏ Calculate parametric arc length.

Chapter 20
- ❏ Apply separation of variables.
- ❏ Find a specific solution to a differential equation.
- ❏ Calculate exponential growth and decay.

Chapter 21
- ❏ Use linear approximation to estimate a function's value.
- ❏ Draw the slope field for a differential equation.
- ❏ Apply Euler's Method to a differential equation.

Chapter 22
- ❏ Define a sequence.
- ❏ Determine the defining rule of a sequence given a list of its elements.
- ❏ Determine the convergence of a sequence.
- ❏ Distinguish between a sequence and a series.
- ❏ Apply the nth term divergence test.
- ❏ Determine convergence and calculate sums of geometric series.
- ❏ Determine convergence of p-series.
- ❏ Determine convergence and sums of telescoping series.

Chapter 23
- ❏ Apply the Integral Test.
- ❏ Apply the Comparison Test.
- ❏ Apply the Limit Comparison Test.
- ❏ Apply the Ratio Test.
- ❏ Apply the Root Test.
- ❏ Apply the Alternating Series Test.
- ❏ Determine whether or not a series converges absolutely.

Chapter 24
- ❏ Find the radius and interval of convergence for a power series.
- ❏ Design and use a Maclaurin series of degree n to estimate a function's values.
- ❏ Design and use a Taylor series of degree n to estimate a function's values.

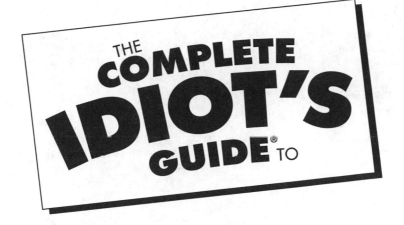

THE **COMPLETE IDIOT'S GUIDE** TO

Calculus

by W. Michael Kelley

ALPHA

A Pearson Education Company

To Lisa, who shares my heart, my home, and my hamburgers. Had I a thousand years to find one as special as you, it would still not be enough. You are my love and the undisputed queen of the Roger Rabbit dance.

For marketing and publicity, please call: 317-581-3722

The publisher offers discounts on this book when ordered in quantity for bulk purchases and special sales.

For sales within the United States, please contact: Corporate and Government Sales, 1-800-382-3419 or corpsales@pearsontechgroup.com

Outside the United States, please contact: International Sales, 317-581-3793 or international@pearsontechgroup.com

Publisher: *Marie Butler-Knight*
Product Manager: *Phil Kitchel*
Managing Editor: *Jennifer Chisholm*
Acquisitions Editor: *Mike Sanders*
Development Editor: *Nancy D. Lewis*
Production Editor: *Billy Fields*
Copy Editor: *Amy Borrelli*
Illustrator: *Chris Eliopoulos*
Cover/Book Designer: *Trina Wurst*
Indexer: *Tonya Heard*
Layout/Proofreading: *Juli Cook, Svetlana Dominguez*

Contents at a Glance

Contents

Appendixes

Foreword

Here's a new one—a Calculus book that doesn't take itself too seriously! I can honestly say that in all my years as a math major, I've never come across a book like this.

My name is Danica McKellar. I am primarily an actress and filmmaker (probably most recognized by my role as "Winnie Cooper" on "The Wonder Years"), but a while back I took a 4-year sidetrack and majored in Mathematics at UCLA. During that time I also co-authored the proof of a new math theorem and became a published mathematician. What can I say? I love math!

But let's face it. You're not buying this book because you love math. And that's okay. Frankly, most people don't love math as much as I do … or at all for that matter. This book is not for the dedicated math majors who want every last technical aspect of each concept explained to them in precise detail.

This book is for every Bio major who has to pass two semesters of Calculus to satisfy the university's requirements. Or for every student who has avoided mathematical formulas like the plague, but is suddenly presented with a whole textbook full of them. I knew a student who switched majors from Chemistry to English, in order to avoid Calculus! Oh, Mr. Kelley, couldn't you have written this book a few years ago?

Mr. Kelley provides explanations that give you the broad strokes of Calculus concepts—and then he follows up with specific tools (and tricks!) to solve some of the everyday problems that you will encounter in your Calculus classes.

You can breath a sigh of relief—the content of this book will not demand of you what your other Calculus textbooks do. I found the explanations in this book to be, by and large, friendly and casual. The definitions don't concern themselves with high-end accuracy, but will bring home the essence of what the heck your textbook was trying to describe with their 50-cent math words. In fact, don't think of this as a textbook at all. What you will find here is a conversation on paper that will hold your hand, make jokes(!), and introduce you to the major topics you'll be required to learn for your current Calculus class. The friendly tone of this book is a welcome break from the clinical nature of every other math book I've ever read!

And oh, Mr. Kelley's colorful metaphors—comparing piecewise functions to Frankenstein's body parts—well, you'll understand when you get there ☺.

My advice would be to read the chapters of *this* book as a nonthreatening introduction to the basic Calculus concepts, and then for fine-tuning, revisit your class's textbook. Your textbook explanations should make much more sense after reading this book, and you'll be more confident and much better qualified to appreciate the specific details required of you by your class. Then you can remain in control of how detailed and nit-picky you want to be in terms of the *mathematical* precision of your understanding by consulting your "unfriendly" Calculus textbook.

Congratulations for taking on the noble pursuit of Calculus! And even more congratulations to you for being proactive and buying this book. As a supplement to your more rigorous textbook, you won't find a friendlier companion.

Good luck!

Danica McKellar

Actress, Summa cum laude, Bachelor of Science in Pure Mathematics at UCLA

Introduction

Let's be honest. Most people would like to learn calculus as much as they'd like to be kicked in the face by a mule. Usually, they have to take the course because it's required or they walked too close to the mule, in that order. Calculus is dull, calculus is boring, and calculus didn't even get you anything for your birthday.

It's not like you didn't try to understand calculus. You even got this bright idea to try and read your calculus textbook. What a joke that was. You're more likely to receive the Nobel Prize for Chemistry than to understand a single word of it. Maybe you even asked a friend of yours to help you, and talking to her was like trying to communicate with an Australian aborigine. You guys just didn't speak the same language.

You wish someone would explain things to you in a language that you understand, but in the back of your mind, you know that the math lingo is going to come back to haunt you. You're going to have to understand it in order to pass this course, and you don't think you've got it in you. Guess what? You do!

Here's the thing about calculus: Things are never as bad as they seem. The mule didn't mean it, and I know this great plastic surgeon. I also know how terrifying calculus is. The only thing scarier than learning it is teaching it to 35 high school students in a hot, crowded room right before lunch. I've fought in the trenches at the front line and survived to tell the tale. I can even tell it in a way that may intrigue, entertain, and teach you something along the way.

We're going to journey together for a while. Allow me to be your guide in the wilderness that is calculus. I've been here before and I know the way around. My goal is to teach you all you'll need to know to survive out here on your own. I'll explain everything in plain and understandable English. Whenever I work out a problem, I'll show you every step (even the simple ones) and I'll tell you exactly what I'm doing and why. Then you'll get a chance to practice the skill on your own without my guidance. Never fear, though—I answer the question for you fully and completely in the back of the book.

I'm not going to lie to you. You're not going to find every single problem easy, but you will eventually do every one. All you need is a little push in the right direction, and someone who knows how you feel. With all these things in place, you'll have no trouble hoofing it out. Oh, sorry, that's a bad choice of words.

How This Book Is Organized

This book is presented in five parts.

In **Part 1, "The Roots of Calculus,"** you'll learn why calculus is useful and what sorts of skills it adds to your mathematical repertoire. You'll also get a taste of its history, which is marred by quite a bit of controversy. Being a math person, and by no means a history buff, we'll get into the math without much delay. However, before we can actually start discussing calculus concepts, we'll spend some quality time reviewing some prerequisite algebra and trigonometry skills.

In **Part 2, "Laying the Foundation for Calculus,"** it's time to get down and dirty. This is the moment you've been waiting for. Or is it? Most people consider calculus the study of derivatives and integrals, and we don't really talk too much about those two guys until Part 3. Am I just a royal tease? Nah. First, we have to talk about limits and continuity. These foundational concepts constitute the backbone for the rest of calculus, and without them, derivatives and integrals couldn't exist.

Finally, we meet one of the major players in **Part 3, "The Derivative."** The name says it all. All of your major questions will be answered, including what a derivative is, how to find one, and what to do if you run into one in a dark alley late at night. (Run!) You'll also learn a whole slew of major derivative-based skills: drawing graphs of functions you've never seen, calculating how quickly variables change in given functions, and finding limits that once were next to impossible to calculate. But wait, there's more! How could something called a "wiggle graph" be anything but a barrel of giggles?

In **Part 4, "The Integral,"** you meet the other big boy of calculus. Integration is almost the same as differentiation, except that you do it backwards. Intrigued? You'll learn how the area underneath a function is related to this backwards derivative, called an "antiderivative." It's also time to introduce the Fundamental Theorem of Calculus, which (once and for all) describes how all this crazy stuff is related. You'll find out that integrals are a little more disagreeable than derivatives were; they require you to learn more techniques, some of which are extremely interesting and (is it possible?) even a little fun!

Now that you've met the leading actor and actress in this mathematical drama, what could possibly be left? In **Part 5, "Differential Equations, Sequences, and Series,"** you meet the supporting cast. Although they play only very small roles, calculus wouldn't be calculus without them. You'll experiment with differential equations using slope fields and Euler's Method, two techniques that have really gained popularity in the last decade of calculus (and you thought that calculus has been the same since the beginning of time …). Finally, you'll play around with infinite series, which are similar to puzzles you've seen since you started kindergarten ("Can you name the next number in this pattern?").

Things to Help You Out Along the Way

As a teacher, I constantly found myself going off on tangents—everything I mentioned reminded me of something else. These peripheral snippets are captured in this book as well. Here's a guide to the different sidebars you'll see peppering the pages that follow.

Critical Point

These notes, tips, and thoughts will assist, teach, and entertain. They add a little something to the topic at hand, whether it be some sound advice, a bit of wisdom, or just something to lighten the mood a bit.

Talk the Talk

Calculus is chock-full of crazy- and nerdy-sounding **words** and **phrases.** In order to become King or Queen Math Nerd, you'll have to know what they mean!

Kelley's Cautions

Although I will warn you about common pitfalls and dangers throughout the book, the dangers in these boxes deserve special attention. Think of these as skulls and crossbones painted on little signs that stand along your path. Heeding these cautions can sometimes save you hours of frustration.

You've Got Problems

Math is not a spectator sport! Once we discuss a topic, I'll explain how to work out a certain type of problem, and then you have to try it on your own. These problems will be very similar to those that I walk you through in the chapters, but now it's your turn to shine. Even though all the answers appear in Appendix A, you should only look there to check your work.

Acknowledgments

There are many people who supported, cajoled, and endured me when I undertook the daunting task of book writing. Although I cannot thank all those who helped me, I do want to name a few of them here. First of all, thanks to the people who made this book possible: Jessica Faust (for tracking me down and getting me to write this puppy), Mike Sanders (who gave the green light), Nancy Lewis (who is the only person on earth who actually *had* to read this *whole thing*), and Sue Strickland (who reviewed for technical accuracy because she supports me no matter what I do, and who'd have done it standing on her head if it reduced my stress level). On a more personal level, there are a few other people I need to thank.

To Lisa, who makes my life better and easier, by just being herself. No man could ask for more than a warm hug and a treacherously placed shoe trap. I'm not an easy man to live with, and no one knows that better than I. Thanks for your patience, your kindness, and always telling me where the salad spinner goes, since Lord knows I'll never remember.

To Mom and Dad. You taught me a lot, even when you didn't intend it. It's now for me to see what I do with it.

To Dave, the Dawg (also spelled D-O-double G). Can you have a better friend than a brother? Here's hoping that Philly makes it to the big dance one day.

To the former students who have chosen to keep in touch with their weird math teacher. I count you among my greatest accomplishments, even if I can take no credit for how great you guys are. I especially need to thank Jason "Supervillain" B., Terumi "Thumper" C., James F., Tim F., Blake H., Laura "Ernie" K., Melissa M., Karen "Pinkie" O., Becky "Patch" R., Emily "Hercules" R., Lori S., Katie and Laura S., and Jon "Doc" W.

To Rob, Chris, and Matt: Three great guys with whom I have shared very, very, very squalid apartments. How many people could stay friends after noting things like "I think that rice has been on the floor for eight months," "I never thought I'd see a cucumber completely liquefied," and "When I finally moved the dishes in the sink I heard this gross popping sound." Truly, if we are not blood brothers, we are crud brothers.

Finally, to Joe, who always asked how the book was going, and for assuring me it'd be a "home run."

Special Thanks to the Technical Reviewer from the Publisher

The Complete Idiot's Guide to Calculus was reviewed by Susan Strickland, an expert who double-checked the accuracy of what you'll learn here. The publisher would like to extend our thanks to Sue in helping us ensure that this book gets all its facts straight.

Susan Strickland received a B.S. in Mathematics from St. Mary's College of Maryland in 1979, an M.S. in Mathematics from Lehigh University in Bethlehem, Pennsylvania, in 1982, and took graduate courses in Mathematics Education at The American University in Washington, D.C., from 1989 through 1991. She was an assistant professor of mathematics and supervised student mathematics teachers at St. Mary's College of Maryland from 1983 through 2001. In the summer of 2001, she accepted the position as a professor of mathematics at the College of Southern Maryland, where she expects to be until she retires! Her interests include teaching mathematics to the "math phobics," training new math teachers, and solving math games and puzzles.

Trademarks

Part 1

The Roots of Calculus

You've heard of Newton, haven't you? If not the man, then at least the fruit-filled cookie? Well, the Sir Isaac variety of Newton is one of the two men responsible for bringing calculus into your life and your course-requirement list. Actually, he is just one of the two men that should shoulder the blame. Calculus's history is long, however, and its concepts predate either man. Before we start studying calculus, we'll take a (very brief) look at its history and development and answer that sticky question: "Why do I have to learn this?"

Next, it's off to practice our prerequisite math skills. You wouldn't try to bench-press 300 pounds without warming up first, would you? A quick review of linear equations, factoring, quadratic equations, function properties, and trigonometry will do a body good. Even if you think you're ready to jump right into calculus, this brief review is recommended. I bet you've forgotten a few things you'll need to know later, so take care of that now!

What Is Calculus, Anyway?

In This Chapter

- ◆ Why calculus is useful
- ◆ The historic origins of calculus
- ◆ The authorship controversy
- ◆ Can I ever learn this?

The word *calculus* can mean one of two things: a computational method or a mineral growth in a hollow organ of the body, such as a kidney stone. Either definition often personifies the pain and anguish endured by students trying to understand the subject. It is far from controversial to suggest that mathematics is not the most popular of subjects in contemporary education, but calculus holds the great distinction of King of the Evil Math Realm, especially by the math phobic. It represents an unattainable goal, an unthinkable miasma of confusion and complication, and few venture into its realm unless propelled by such forces as job advancement or degree requirement. No one knows how much people fear calculus more than a calculus teacher.

The minute people find out that I taught a calculus class, they are compelled to describe, in great detail, exactly how they did in high school math, what subject they "topped out" in, and why they feel that calculus is the embodiment of evil. Most of these people are my barbers, and I can't explain why. All of the friendly folks at the Hair Cuttery have come to know me as the strange balding man with arcane and baffling mathematical knowledge.

Most of the fears surrounding calculus are unjustified. Calculus is a step up from high school algebra, no more. Following a straightforward list of steps, just like you do with most algebra problems, solves the majority of calculus problems. Don't get me wrong—calculus is not always easy, and the problems are not always trivial, but it is not as imposing as it seems. Calculus is a truly fascinating tool with innumerable applications to "real life," and for those of you who like soap operas, it's got one of the biggest controversies in history to its credit.

What's the Purpose of Calculus?

Calculus is a very versatile and useful tool, not a one-trick pony by any stretch of the imagination. Many of its applications are direct upgrades from the world of algebra—methods of accomplishing similar goals, but in a far greater number of situations.

Critical Point

What we call "calculus," scholars call "*the* calculus." Because any method of computation can be called a calculus and the discoveries comprising modern-day calculus are so important, the distinction is made to clarify. I personally find the terminology a little pretentious and won't use it. I've never been asked "Which calculus are you talking about?"

Whereas it would be impossible to list all the uses of calculus, the following list represents some interesting highlights of the things you will learn by the end of the book.

Finding the Slopes of Curves

One of the earliest algebra topics learned is how to find the slope of a line—a numerical value that describes just how slanted that line is. Calculus affords us a much more generalized method of finding slopes. With it, we can find not only how steeply a line slopes, but indeed, how steeply any curve slopes at any given time. This might not at first seem useful, but it is actually one of the most handy mathematics applications around.

Calculating the Area of Bizarre Shapes

Without calculus, it is difficult to find areas of shapes other than those whose formulas you learned in geometry. Sure, you may be a pro at finding the area of a circle, square, rectangle, or triangle, but how would you find the area of a shape like the one shown in Figure 1.1?

Justifying Old Formulas

There was a time in your math career that you took formulas on faith. Sometimes we still need to do that, but calculus affords us the opportunity to finally verify some of those old formulas, especially from geometry. You were always told that the volume of a cone was one third the volume of a cylinder with the same Vadius $\left(V = \frac{1}{3}\pi r^2 h\right)$, but through a simple calculus process of three-dimensional linear rotation, we can finally prove it. (By the way, the process is simple even though it may not sound like it right now.)

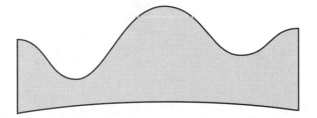

Figure 1.1

Calculate this area? We're certainly not in Kansas anymore ....

Calculate Complicated *X*-Intercepts

Without the aid of a graphing calculator, it is exceptionally hard to calculate an *irrational root*. However, a simple, repetitive process called Newton's Method (named after Sir Isaac Newton) allows you to calculate an irrational root to whatever degree of accuracy you desire.

Visualizing Graphs

You may already have a good grasp of lines and how to visualize their graphs easily, but what about the graph of something like $y = x^3 + 2x^2 - x + 1$?

Talk the Talk

An **irrational root** is an *x*-intercept that is not a fraction. Fractional (rational) roots are much easier to find, because you can typically factor the expression to calculate them, a process that is taught in the earliest algebra classes. No good, generic process of finding irrational roots is possible until you use calculus.

Very elementary calculus tells you exactly where that graph will be increasing, decreasing, and twisting. In fact, you can find the highest and lowest points on the graph without plotting a single point.

Finding the Average Value of a Function

Anyone can average a set of numbers, given the time and the fervent desire to divide. Calculus allows you to take your averaging skills to an entirely new level. Now you can even find, on average, what height a function travels over a period of time. For example, if you graph the path of an airplane (see Figure 1.2), you can calculate its average cruising altitude with little or no effort. Determining its average velocity and acceleration are no harder. You may never have had the impetus to do such a thing, but you've got to admit that it's certainly more interesting than averaging the odd numbers less than 50.

Figure 1.2

Even though this plane's flight path is not defined by a simple shape (like a semi-circle), using calculus you can calculate all sorts of things, like its average altitude during the journey or the number of complementary peanuts you dropped when you fell asleep.

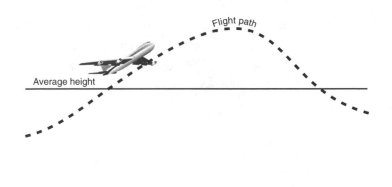

Calculating Optimal Values

One of the most mind-bendingly useful applications of calculus is the optimization of functions. In just a few steps, you can answer questions such as: "If I have 1,000 feet of fence, what is the largest rectangular yard I can make?" or "Given a rectangular sheet of paper which measures 8.5 inches by 11 inches, what are the dimensions of the box I can make containing the greatest volume?" The traditional way to create an open box from a rectangular surface is to cut congruent squares from the corners of the rectangle and then to fold the resulting sides up, as shown in Figure 1.3.

Figure 1.3

With a few folds and cuts, you can easily create an open box from a rectangular surface.

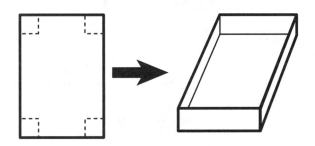

I tend to think of learning calculus and all of its applications as suddenly growing a third arm. Sure, it may feel funny having a third arm at first. In fact, it'll probably make you stand out in bizarre ways from those around you. However, given time, you're sure to find many uses for that arm that you'd have never imagined without having first possessed it.

Who's Responsible for This?

Tracking the discovery of calculus is not as easy as, say, tracking the discovery of the safety pin. Any new mathematical concept is usually the result of hundreds of years of investigation, debate, and debacle. Many come close to stumbling upon key concepts, but only the lucky few who finally make the small, key connections receive the credit. Such is the case with calculus.

Calculus is usually defined as the combination of the differential and integral techniques you will learn later in the book. However, historical mathematicians would never have swallowed the concepts we take for granted today. The key ingredient missing in mathematical antiquity was the hairy notion of infinity. Mathematicians and philosophers of the time had an extremely hard time conceptualizing infinitely small or large quantities. Take, for instance, the Greek philosopher Zeno.

Ancient Influences

Zeno took a very controversial position in mathematical philosophy: He argued that all motion is impossible. In the paradox titled Dichotomy, he used a compelling, if not strange, argument illustrated in Figure 1.4.

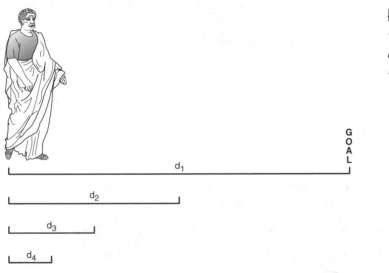

Figure 1.4

The infinite subdivisions described in Zeno's Dichotomy.

Critical Point

The most famous of Zeno's paradoxes is a race between a tortoise and the legendary Achilles called, appropriately, *the Achilles*. Zeno contends that if the tortoise has a head start, no matter how small, Achilles will never be able to close the distance. To do so, he'd have to travel half of the distance separating them, then half of that, ad nauseum, presenting the same dilemma illustrated by the Dichotomy.

In Zeno's argument, the individual pictured wants to travel to the right, to his eventual destination. However, before he can travel that distance (d_1), he must first travel half of that distance (d_2). That makes sense, since d_2 is smaller and comes first in the path. However, before the d_2 distance can be completed, he must first travel half of *it* (d_3). This procedure can be repeated indefinitely, which means that our beleaguered sojourner must travel an infinite number of distances. No one can possibly do an infinite number of things in a finite amount of time, says Zeno, since an infinite list will never be exhausted. Therefore, not only will the man never reach his destination, he will, in fact, never start moving at all! This could account for the fact that you never seem to get anything done on Friday afternoons.

Critical Point

In case the suspense is killing you, let me ruin the ending for you. The essential link to completing calculus and satisfying everyone's concerns about infinite behavior was the concept of limit, which laid the foundation for both derivatives and integrals.

Zeno didn't actually believe that motion was impossible. He just enjoyed challenging the theories of his contemporaries. What he, and the Greeks of his time, lacked was a good understanding of infinite behavior. It was unfathomable that an innumerable number of things could fit into a measured, fixed space. Today, geometry students accept that a line segment, though possessing fixed length, contains an infinite number of points. The development of some reasonable and yet mathematically sound concept of very large quantities or very small quantities was required before calculus could sprout.

Some ancient mathematicians weren't troubled by the apparent contradiction of an infinite amount in a finite space. Most notably, Euclid and Archimedes contrived the method of exhaustion as a technique to finding the area of a circle, since the exact value of π wouldn't be around for some time. In this technique, regular polygons were inscribed in a circle; the higher the number of sides of the polygon, the closer the area of the polygon would be to the area of the circle (see Figure 1.5).

Figure 1.5

The higher the number of sides, the closer the area of the inscribed polygon approximates the area of the circle.

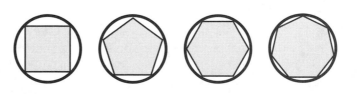

In order for the method of exhaustion (which is aptly titled, in my opinion) to give the exact value for the circle, the polygon would have to have an infinite number of sides. Indeed, this magical incarnation of geometry can only be considered theoretically, and the idea that a shape of infinite sides could have a finite area made most people of the time very antsy. However, seasoned calculus students of today can see this as a simple limit problem. As the number of sides approaches infinity, the area of the polygon approaches πr^2, where r is the radius of the circle. Limits are essential to the development of both the derivative and integral, the two fundamental components of calculus. Although Newton and Leibniz were unearthing the major discoveries of calculus in the late 1600s and early 1700s, no one had established a formal limit definition. Although this may not keep *us* up at night, it was, at the least, troubling at the time. Mathematicians worldwide started sleeping more soundly at night circa 1751, when Jean Le Rond d'Alembert wrote *Encyclopédie* and established the formal definition of the limit. The delta-epsilon definition of the limit we use today is very close to that of d'Alembert.

Even before its definition was established, however, Newton had given a good enough shot at it that calculus was already taking shape.

Newton vs. Leibniz

Sir Isaac Newton, who was born in poor health in 1642 but became a world-renowned smart guy (even during his own time), once retorted, "If I have seen farther than Descartes, it is because I have stood on the shoulders of giants." No truer thing could be said about any major mathematical discovery, but let's not give the guy too much credit for his supposed modesty … more to come on that in a bit. Newton realized that infinite series (e.g., the method of exhaustion) were not only great approximators, but if allowed to actually reach infinity, they gave the exact values of the functions they approximated. Therefore, they behaved according to easily definable laws and restrictions usually only applied to known functions. Most importantly, he was the first person to recognize and utilize the inverse relationship between the slope of a curve and the area beneath it.

That inverse relationship (contemporarily called the Fundamental Theorem of Calculus) marks Newton as the inventor of calculus. He published his findings, and his intuitive definition of a limit, in his 1687 masterwork entitled *Philosophiae Naturalis Principia Mathematica*. The *Principia*, as it is more commonly known today, is considered by some (those who consider such things, I suppose) to be the greatest scientific work of all time, excepting of course any books yet to be written by the comedian Sinbad. Calculus was actively used to solve the major scientific dilemmas of the time:

- Calculating the slope of the tangent line to a curve at any point along its length
- Determining the velocity and acceleration of an object given a function describing its position, and designing such a position function given the object's velocity or acceleration
- Calculating arc lengths and the volume and surface area of solids
- Calculating the relative and absolute *extrema* of objects, especially projectiles

Talk the Talk

Extrema points are high or low points of a curve (maxima or minima, respectively). In other words, they represent extreme values of the graph, whether extremely high or extremely low, in relation to the points surrounding them.

Critical Point

Ten years after Leibniz's death, Newton erased the reference to Leibniz from the third edition of the *Principia* as a final insult. This is approximately the academic equivalent of Newton throwing a chair at Leibniz on *The Jerry Springer Show* (topic: "You published your solution to an ancient mathematical riddle before me and I'm fightin' mad!").

However, with a great discovery often comes great controversy, and such is the case with calculus.

Enter Gottfried Wilhelm Leibniz, child prodigy and mathematical genius. Leibniz was born in 1646 and completed college, earning his Bachelor's degree, at the ripe old age of 17. Because Leibniz was primarily self-taught in the field of mathematics, he often discovered important mathematical concepts on his own, long after someone else had already published them. Newton actually credited Leibniz in his *Principia* for developing a method similar to his. That similar method evolved into a near match of Newton's work in calculus, and in fact, Leibniz published his breakthrough work inventing calculus *before* Newton, although Newton had already made the exact discovery years before Leibniz. Some argue that Newton possessed extreme sensitivity to criticism and was, therefore, slow to publish. The mathematical war was on: Who invented calculus first and thus deserved the credit for solving a riddle thousands of years old?

Today, Newton is credited for inventing calculus first, although Leibniz is credited for its first publication. In addition, the shadow of plagiarism and doubt has been lifted from Leibniz, and it is believed that he discovered calculus completely independent of Newton. However, two distinct factions arose and fought a bitter war of words. British mathematicians sided with Newton, whereas continental Europe supported Leibniz, and the war was long and hard. In fact, British mathematicians were effectively alienated from the rest of the European mathematical community because of the rift, which probably accounts for the fact that there were no great mathematical discoveries made in Britain for some time thereafter.

Although Leibniz just missed out on the discovery of calculus, many of his contributions live on in the language and symbols of mathematics. In algebra, he was the first to use a dot to indicate multiplication ($3 \cdot 4 = 12$) and a colon to designate a proportion ($1:2 = 3:6$). In geometry, he contributed the symbols for congruent ($\cong$) and similar ($\sim$). Most famous of all, however, are the symbols for the derivative and the integral, which we also use.

Will I Ever Learn This?

History aside, calculus is an overwhelming topic to approach from a student's perspective. There are an incredible number of topics, some of which are related, but most of which are not in any obvious sense. However, there is no topic in calculus that is, in and of itself, very difficult once you understand what is expected of you. The real trick is to quickly recognize what sort of problem is being presented and then to attack it using the methods you will read and learn in this book.

I have taught calculus for a number of years, to high school students and adults alike, and I believe that there are four basic steps to being successful in calculus:

Critical Point

Leibniz also coined the term *function*, which is commonly learned in an elementary algebra class. However, most of Leibniz's discoveries and innovations were eclipsed by Newton, who made great strides in the topics of gravity, motion, and optics (among other things). The two men were bitter rivals and were fiercely competitive against each other.

◆ *Make sure to understand what the major vocabulary words mean.* This book will present all important vocabulary terms in simple English, so you not only understand what the terms mean, but how they apply to the rest of your knowledge.

◆ *Sift through the complicated wording of the important calculus theorems and strip away the complicated language.* Math is just as foreign a language as French or Spanish to someone who doesn't enjoy numbers, but that doesn't mean you can't understand complicated mathematical theorems. I will translate every theorem into plain English and make all the underlying implications perfectly clear.

◆ *Develop a mathematical instinct.* As you read, I will help you recognize subtle clues presented by calculus problems. Most problems do everything but tell you exactly how they must be solved. If you read carefully, you will develop an instinct, a feeling that will tingle in your inner fiber and guide you toward the right answers. This comes with practice, practice, practice, so the majority of the book presents sample problems with detailed solutions to help you navigate the muddy waters of calculus.

◆ *Sometimes you just have to memorize.* There are some very advanced topics covered in calculus that are hard to prove. In fact, many theorems cannot be proven until you take much more advanced math courses. Whenever I think that proving a theorem will help you understand it better, I will do so and discuss it in detail. However, if a formula, rule, or theorem has a proof that I deem unimportant to you mastering the topic in question, I will omit it, and you'll just have to trust me that it's for the best.

The Least You Need to Know

◆ Calculus is the culmination of algebra, geometry, and trigonometry.

◆ Calculus as a tool enables us to achieve greater feats than the mathematics courses that precede it.

◆ Limits are foundational to calculus.

◆ Newton and Leibniz both discovered calculus independently, though Newton discovered it first.

◆ With time and dedication, anyone can be a successful calculus student.

2

Polish Up Your Algebra Skills

In This Chapter

◆ Creating linear equations

◆ The properties of exponents

◆ Factoring polynomials

◆ Solving quadratic equations

If you are an aspiring calculus student, somewhere in your past you probably had to do battle with the beast called algebra. Not many people have positive memories associated with their algebraic experiences, and I am no different. Forget the fact that I was a math major, a calculus teacher, and even took my calculator to bed with me when I was young (a true but very sad story). I hated algebra for many reasons, not the least of which was that I felt I could never keep up with it. Every time I seemed to understand algebra, we'd be moving on to a new topic much harder than the last.

Being an algebra student is just like fighting Mike Tyson. Here is this champion of mathematical reasoning that has stood unchallenged for hundreds of years, and you're in the ring going toe-to-toe with it. You never really reach back for that knockout punch because you're too busy fending off your opponent's blows. When the bell rings to signal the end of the fight, all you can think is "I survived!" and hope that someone can carry you out of the ring.

Perhaps you didn't hate algebra as much as I did. You might be one of those lucky people who understood algebra easily. You are very lucky. For the rest of us, however, there is hope. Algebra is much easier in retrospect than when you were first being pummeled by it. As calculus is a grand extension of algebra, you will, of course, need a large repertoire of algebra skills. So it's time to slip those old boxing gloves back on and go a few rounds with your old sparring partner. The good news is you've undoubtedly gotten stronger since the last bout.

Walk the Line: Linear Equations

Graphs play a large role in calculus, and the simplest of graphs, the line, surprisingly pops up all the time. As such, it is important that you can recognize, write, and analyze graphs and equations of lines. To begin, remember that a line's equation always has three components: Two variables terms and a constant (numeric) term. One of the most common ways to write an equation is in standard form.

Common Forms of Linear Equations

A line in standard form looks like this: $Ax + By = C$. In other words, the variable terms are on the left side and the number is on the right side of the equal sign. Also, to officially be in standard form, the coefficients (A, B, and C) must be *integers*, and A is supposed to be positive. What's the purpose of standard form? A linear equation can have many different forms (for example, $x + y = 2$ is the same line as $x = 2 - y$). However, once in standard form, all lines with the same graph have the exact same equation. Therefore, standard form is especially handy for instructors; they'll often ask that answers be put into standard form to avoid alternate correct answers.

Talk the Talk

An **integer** is a number without a decimal or fractional part. For example, 3 and –6 are integers, whereas 10.3 and $-\frac{1}{2}$ are not.

You've Got Problems

Problem 1: Put the following linear equation into standard form:
$$3x - 4y - 1 = 9x + 5y - 12$$

There are two major ways to create the equation of a line. One requires that you have the slope and the *y*-intercept of the line. Appropriately enough, it is called slope-intercept form: $y = mx + b$. In this equation, m represents the slope and b the *y*-intercept. Notice the major characteristic of an equation in slope-intercept form: It is solved for *y*. In other words, *y* appears by itself on the left side of the equation.

Example 1: Write the equation of a line with slope −3 and y-intercept 5.

Solution: In slope intercept form, $m = -3$ and $b = 5$, so the answer is …

$$y = -3x + 5$$

Another way to create a linear equation requires a little less information—only a point and the slope (the point doesn't have to be the y-intercept). This (thanks to the vast creativity of mathematicians) is called point-slope form. Given the point (x_1, y_1) and slope m, the equation of the resulting line will be $y - y_1 = m(x - x_1)$.

You will find this form extremely handy throughout the rest of your travels with calculus, so make sure you understand it. Don't get confused between the x's and x_1's or the y's and the y_1's. The variables with the subscript represent the coordinates of the point you're given. Don't replace the other x and y with anything—these variables are left in your final answer. Watch how easy this is.

Example 2: If a line g contains the point (−5,2) and has slope $-\frac{1}{5}$, what is the equation of g in standard form?

Solution: Because you are given a slope and a point (which is not the y-intercept) you should use point-slope form to create the equation of the line. Therefore, $m = -\frac{1}{5}$, $x_1 = -5$, and $y_1 = 2$. Plug these values into point-slope form and get:

$$y - 2 = -\frac{1}{5}\left(x - (-5)\right)$$
$$y - 2 = -\frac{1}{5}(x + 5)$$

If this equation is supposed to be in standard form, you're not allowed to have any fractions. Remember that the coefficients have to be integers, so to get rid of the fractions, multiply the entire equation by 5:

$$5y - 10 = -(x + 5)$$
$$5y - 10 = -x - 5$$

Now, move the variables to the left and the constants to the right and make sure the x term is positive; this puts everything in standard form:

$$x + 5y = 5$$

Problem 2: Find the equation of the line through point $(0,-2)$ with slope $\frac{2}{3}$ and put it in standard form.

Calculating Slope

You might have noticed that both of the ways we use to create lines absolutely require that you know the slope of the line. The slope of the line is *that* important (almost as important as wearing both shoes and a shirt if you want to buy a Slurpee at 7-Eleven). The *slope* of a line is a number that describes precisely how "slanty" that line is—the larger the value of the slope, the steeper the line. Furthermore, the sign of the slope (in most cases Capricorn) will tell you whether or not the line rises or falls as it travels.

As shown in Figure 2.1, lines with shallower inclines have smaller slopes. If the line rises (from left to right), the slope is positive; if, however, it falls from left to right, the slope is negative. Horizontal lines have 0 slope (not positive or negative), and vertical lines are said to have an undefined slope, or no slope at all.

Figure 2.1

Calculating the slope of a line.

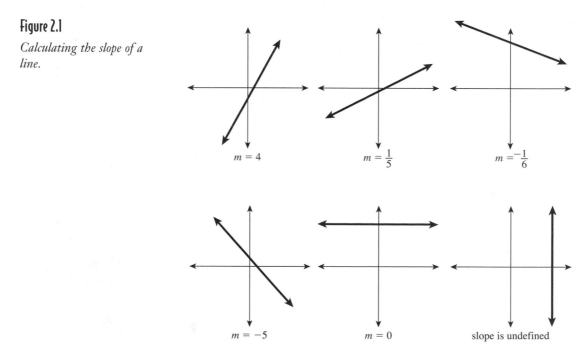

It is very easy to calculate the slope of any line: Find any two points on the line (a,b) and (c,d) and plug them into this formula:

$$\text{slope} = \frac{d - b}{c - a}$$

In essence, you are finding the difference in the y's and dividing by the difference in the x's. If the numerator is larger, the y's are changing faster, and the line is getting steeper. On the other hand, if the denominator is larger, the line is moving more quickly to the left or right than up and down, creating a shallow incline.

You've Got Problems

Problem 3: Find the slope of the line that contains points (3,7) and (–1,4).

You should also remember that parallel lines have equal slopes, whereas perpendicular lines have slopes that are negative reciprocals of one another. Therefore, if line g has slope $\frac{5}{7}$, then a parallel line h would have slope $\frac{5}{7}$ also; a perpendicular line k would have slope $-\frac{7}{5}$. We use this information in the next example.

Example 3: Find the equation of line j given that it is parallel to the line $2x - y = 6$ and contains the point (–1,1).

Solution: This problem requires you to create the equation of a line, and you'll find that the best way to do this every time is via point-slope form. So, you need a point and a slope. Well, you already have the point: (–1,1). Using our keen sense of deduction, we know that only the slope is left to find and that'll be that. But how to find the slope? If j is *parallel* to $2x - y = 6$, then they must have the same slope, so what's the slope of that line? Here's the key: If you solve it for y, it will be in slope-intercept form, and the slope, m, is simply the coefficient of x. When you do so, you get $y = 2x - 6$. Therefore, the slope of both lines is 2, and you can use point-slope form to write the equation of j:

$$y - y_1 = m\left(x - x_1\right)$$
$$y - 1 = 2\left(x - (-1)\right)$$
$$y - 1 = 2\left(x + 1\right)$$

You can stop there, since the problem doesn't require you to put your answer in any specific form.

You've Got the Power: Exponential Rules

I find that exponents are the bane of many calculus students. Whether they never learned exponents well in the first place or simply make careless mistakes, exponential errors are a treasure trove of frustration. Therefore, it's worth your while to spend a few minutes and refresh yourself on the major exponential rules. You may find this exercise "empowering." If so, call and tell Oprah, because it might earn me a guest spot on her show.

♦ Rule one: $x^a \cdot x^b = x^{a+b}$

Explanation: If you multiply two terms with the same base (here it's x), add the powers and keep the base. For example, $a^2 \cdot a^7 = a^9$.

♦ Rule two: $\dfrac{x^a}{x^b} = x^{a-b}$

Explanation: This is the opposite of rule one. If you divide (instead of multiply) two terms with the same base, then you subtract (instead of add) the powers and keep the base. For example, $\dfrac{w^7}{w^3} = w^4$.

♦ Rule three: $x^{-a} = \dfrac{1}{x^a}$

Explanation: A negative exponent indicates that a variable is in the wrong spot, and belongs in the opposite part of the fraction, but it only affects the variable it's touching. For example, in the expression $\dfrac{x^3 y^{-2}}{3}$, only the y is raised to a negative power, so it needs to be in the opposite part of the fraction. Correctly simplified, that fraction looks like this: $\dfrac{x^3}{3y^2}$. Note that the exponent becomes positive when it moves to the right place. Remember that a happy (positive) exponent is where it belongs in a fraction.

Critical Point

Eliminate negative exponents in your answers. Most instructors consider an answer with negative exponents in it unsimplified. They must see the glass as half-empty. Think about it. How many cheery math teachers do *you* know?

♦ Rule four: $(x^a)^b = x^{ab}$

Explanation: If an exponential expression is raised to a power, you should multiply the exponents and keep the base. For example, $(b^7)^3 = b^{21}$.

You've Got Problems

Problem 4: Simplify the expression $(3x^{-3}y^2)^2$ using exponential rules.

♦ Rule five: $x^{a/b} = \sqrt[b]{x^a}$ and $\left(\sqrt[b]{x}\right)^a$

Explanation: The numerator of the fractional power remains the exponent. The denominator of the power tells you what sort of radical (square root, cube root, etc.). For example, $4^{3/2}$ can be simplified as either $\sqrt{4^3}$ or $\left(\sqrt{4}\right)^3$. Either way, the answer is 8.

Example 4: Simplify $xy^{1/3}(x^2y)^3$.

Solution: Your first step should be to raise (x^2y) to the third power. You have to use rule four twice (the current exponent of y is understood to be 1 if it is not written). This gives you $x^{2\cdot3}y^{1\cdot3} = x^6y^3$. The problem now looks like: $xy^{1/3}\left(x^6y^3\right)$.

To finish, you have to multiply the x's and y's together using rule one:

$$x \cdot x^6 \cdot y^{1/3} \cdot y^3$$
$$x^{1+6}y^{3+\frac{1}{3}}$$
$$x^7y^{10/3}$$

Breaking Up Is Hard to Do: Factoring Polynomials

Factoring is one of those things you see over and over and over again in algebra. I have found that even among my students that disliked math, factoring was popular … it's something that some people just "got," even when most of everything else escaped them. This is not the case, however, in many European schools, a fact that surprised my colleagues and me when I was a high school teacher.

Even Canadian exchange students gave me blank stares when we discussed factoring in class. This is not to say that these students were not extremely intelligent (which they were); they just used other methods. However, factoring does come in very handy throughout calculus, so I deem it important enough to earn it some time here. Call it patriotism.

Talk the Talk

Factoring is the process of "unmultiplying," breaking a number or expression down into parts that, if multiplied together, return the original quantity.

Calculus does not require that you factor complicated things, so we'll stick to the basics here. *Factoring* is basically reverse multiplying—undoing the process of multiplication to see what was there to begin with. For example, you can break down the number 6 into factors of 3 and 2, since $3 \cdot 2 = 6$. There can be more than one correct way to factor something.

Greatest Common Factors

Factoring using the greatest common factor is the easiest method of factoring and is used whenever you see terms that have pieces in common. This is much easier than it sounds. Take, for example, the expression $4x + 8$.

Notice that both terms can be divided by 4, making 4 a common factor. Therefore, you can write the expression in the factored form of $4(x + 2)$.

In effect, I have "pulled out" the common factor of 4, and what's left behind are the terms once 4 has been divided out of each. In these type of problems, you should ask yourself, "What do each of the terms have in common?" and then pull that greatest common factor out of each to write your answer in factored form.

> ### You've Got Problems
>
> Problem 5: Factor the expression $7x^2y - 21xy^3$.

Special Factoring Patterns

You should feel comfortable factoring trinomials such as $x^2 + 5x + 4$ using whatever method suits you. Most people play with binomial pairs until they stumble across something that works, in this case $(x + 4)(x + 1)$, whereas others undergo more complicated means. Regardless of your personal "flair," there are some patterns that you should have memorized:

◆ Difference of perfect squares: $a^2 - b^2 = (a + b)(a - b)$

Explanation: A perfect square is a number like 16, which can be created by multiplying something times itself. In the case of 16, that something is 4, since 4 times itself is 16. If you see one perfect square being subtracted from another, you can automatically factor it using the pattern above. For example, $x^2 - 25$ is a difference of x^2 and 25, and both are perfect squares. Thus, it can be factored as $(x + 5)(x - 5)$.

◆ Sum of perfect cubes: $a^3 + b^3 = (a + b)(a^2 - ab + b^2)$

Kelley's Cautions

You cannot factor the *sum* of perfect squares, so whereas $x^2 - 4$ is factorable, $x^2 + 4$ is not!

Explanation: Perfect cubes are similar to perfect squares. The number 125 is a perfect cube because $5 \cdot 5 \cdot 5 = 125$. This pattern is a little clumsier to memorize, but it can be handy occasionally. This formula can be altered just slightly to factor the *difference* of perfect cubes, as illustrated in the next bullet. Other than a couple of sign changes, the process is the same.

◆ Difference of perfect cubes: $a^3 - b^3 = (a - b)(a^2 + ab + b^2)$

Explanation: Enough with the symbols for these formulas—let's do an example.

Example 5: Factor $x^3 - 27$ using the difference of perfect cubes factoring pattern.

Solution: Note that x is a perfect cube since $x \cdot x \cdot x = x^3$, and 27 is also, since $3 \cdot 3 \cdot 3 = 27$. Therefore, $x^3 - 27$ corresponds to $a^3 - b^3$ in the previous formula, making $a = x$ and $b = 3$. Now, all that's left to do is plug a and b into the formula:

$$(a - b)\left(a^2 + ab + b^2\right)$$
$$(x - 3)\left(x^2 + 3x + 9\right)$$

You cannot factor $(x^2 + 3x + 9)$ any further, so you are finished.

You've Got Problems

Problem 6: Factor the expression $8x^3 + 343$.

Solving Quadratic Equations

Before you put algebra review in the rearview mirror, there's one last stop. Sure, you've been able to solve equations like $x + 9 = 12$ since you can remember, but when the equations get a little trickier, maybe you get a little panicky. Forgetting how to solve quadratic equations (equations whose highest exponent is a 2) has distinct symptoms: dizziness, shortness of breath, nausea, and loss of appetite. To fight this ailment, take the following 3 tablespoons of quadratic problem solving and call me in the morning.

Every quadratic equation can be solved with the *quadratic formula* (method three below), but it's important that you know the other two methods as well. Factoring is undoubtedly the fastest of the three methods, so you should definitely try it first. Few people choose completing the square as their first option, but it (like the quadratic formula) works every time, though it has a few more steps than its counterpart. However, you *have* to learn completing the square, because it pops up later in calculus, when you least expect it.

Method One: Factoring

To begin, set your quadratic equation equal to 0; this means add and subtract the terms as necessary to get them all to one side of the equation. If the resulting equation is factorable, factor it and set each individual term equal to 0. These little baby equations will give you the solutions to the equation. That's all there is to it.

Example 6: Solve the equation $3x^2 + 4x = -1$ by factoring.

Solution: Always start the factoring method by setting the equation equal to 0. In this case, start by adding 1 to each side of the equation: $3x^2 + 4x + 1 = 0$.

Now, factor the equation and set each factor equal to 0. This creates two cute little mini-equations that need to be solved, giving you the final answer:

$$(3x + 1)(x + 1) = 0$$
$$3x + 1 = 0 \quad and \quad x + 1 = 0$$
$$x = -\frac{1}{3} \quad and \quad x = -1$$

This equation has two solutions: $x = -\frac{1}{3}$ and $x = -1$. You can check them by plugging each separately into the original equation, and you'll find that the equation becomes true.

Method Two: Completing the Square

As I mentioned earlier, this method is a little trickier than the other two, but you really do need to learn it now, or you'll be coming back to figure it out later. I've learned that it's best to learn this method in the context of an example, so let's go to it.

Example 7: Solve the equation $2x^2 + 12x - 18 = 0$ by completing the square.

Solution: In this method, unlike factoring, you want the constant separate from the variable terms, so move the constant to the right side of the equation by adding 18 to both sides:

$$2x^2 + 12x = 18$$

Kelley's Cautions

If you don't make the coefficient of the x^2 term 1, then the rest of the completing the square process will not work. Also, when you divide to eliminate the x^2 coefficient, make sure you divide *every term* in the equation (including the constant, sitting dejectedly on the other side of the equation).

This is important: For completing the square to work, the coefficient of x^2 *must* be 1. In this case, it is 2, so to eliminate that pesky coefficient, divide every term in the equation by 2:

$$x^2 + 6x = 9$$

Here's the key to completing the square: Take half of the coefficient of the x term, square it, and add it to both sides. In this problem, the x coefficient is 6, so take half of it (3) and square that ($3^2 = 9$). Add the result (9) to both sides of the equation:

$$x^2 + 6x + 9 = 9 + 9$$
$$x^2 + 6x + 9 = 18$$

At this point, if you've done everything correctly, the left side of the equation will be factorable. In fact, it will be a perfect square!

$$(x + 3)(x + 3) = 18$$
$$(x + 3)^2 = 18$$

To solve the equation, take the square root of both sides. That will cancel out the exponent. Whenever you do this, you have to add a ± sign in front of the right side of the equation. This is always done when square rooting both sides of any equation:

$$\sqrt{(x + 3)^2} = \pm\sqrt{18}$$
$$x + 3 = \pm\sqrt{18}$$

To solve for x, subtract 3 from each side, and that's it. It would also be good form to simplify $\sqrt{18}$ into $3\sqrt{2}$:

$$x = -3 \pm 3\sqrt{2}$$

Method Three: The Quadratic Formula

The quadratic formula is one-stop shopping for all your quadratic equation needs. All you have to do is make sure your equation is set equal to 0, and you're halfway there. Your equation will then look like this: $ax^2 + bx + c = 0$, where a, b, and c are the coefficients as indicated. Take those numbers and plug them straight into this formula (which you should definitely memorize):

$$x = \frac{-b \pm \sqrt{b^2 - 4ac}}{2a}$$

You'll get the same answer you would when completing the square. In fact, this formula is pretty easy to generate—all you do is complete the square on the equation $ax^2 + bx + c = 0$. Just to convince you that the answer's the same, we'll do the problem in Example 7 again, but this time with the quadratic formula.

Example 8: Solve the equation $2x^2 + 12x - 18 = 0$, this time using the quadratic formula.

Solution: Because the equation is already set equal to 0, it is in form $ax^2 + bx + c = 0$, and $a = 2$, $b = 12$, and $c = -18$. Plug these values into the quadratic formula and simplify:

$$x = \frac{-12 \pm \sqrt{12^2 - 4(2)(-18)}}{2(2)}$$

$$x = \frac{-12 \pm \sqrt{144 - (-144)}}{4}$$

$$x = \frac{-12 \pm \sqrt{288}}{4}$$

$$x = \frac{-12 \pm 12\sqrt{2}}{4}$$

$$x = \frac{4(-3 \pm 3\sqrt{2})}{4}$$

$$x = -3 \pm 3\sqrt{2}$$

So although there are fewer steps to the quadratic formula, there is some room for error during computation. You should practice both methods, but primarily use the one that feels more comfortable to you.

You've Got Problems

Problem 7: Solve the equation $3x^2 + 12x = 0$ three times, using all the methods you have learned for solving quadratic equations.

The Least You Need to Know

◆ Basic equation solving is an important skill in calculus.

◆ Reviewing the five exponential rules will prevent arithmetic mistakes in the long run.

◆ You can create the equation of a line with just a little information using point-slope form.

◆ There are three major ways to solve quadratic equations, each important for different reasons.

Chapter 3

Equations, Relations, and Functions, Oh My!

In This Chapter

- ◆ When is an equation a function?
- ◆ Important function properties
- ◆ Building your function skills repertoire
- ◆ The basics of parametric equations

I still remember the fateful day in Algebra I when the equation $y = 3x + 2$ became $f(x) = 3x + 2$. The dreaded function! At the time, I didn't quite understand why we had to make the switch. I was a fan of the y, and was sad to see it go. What I failed to grasp was that the advent of the function marked a new step forward in my math career.

If you know that an equation is also a function, it guarantees that the equation in question will always behave in a certain way. Most of the definitions in calculus require functions in order to operate correctly. Therefore, the vast majority of our work in calculus will be with functions exclusively, with the exception of parametric equations. Therefore, it's good to know exactly what a function is, to be able to recognize important functions at a glance, and to be able to perform basic function operations.

What Makes a Function Tick?

Let's get a little vocabulary straight before we get too far. Any sort of equation in mathematics is classified as a *relation*, as the equation describes a specific way that the variables and numbers in the equation are related. Relations don't have to be equations, although that is how they are most commonly written.

Talk the Talk

A **relation** is a collection of related numbers. Most often, the relationship between the numbers is described by an equation, although it can be given simply as a list of ordered pairs. A **function** is a relation such that every input has only one matching output. Any function input is a part of that function's **domain**, and any possible output for the function is part of its **range**.

Here's the most basic definition of a relation. You'll notice that there's not a whole lot to it, just a list of ordered pairs:

$$s:\{(-1,5),(1,6),(2,4)\}$$

This relation, called s, gives a list of inputs and outputs. In essence, you're asking s, "What will you give me if I give you –1?" The reply is 5, since the ordered pair (–1,5) appears in the relation. If you input 2, s spits back 4. However, if you input 6, s has no response; the only inputs s accepts are –1, 1, and 2, and the only outputs it can offer are 5, 6, and 4.

In calculus, it is more useful to write relations like this:

$$g(x) = \frac{1}{3}x - 3$$

This relation, called g, accepts any real number input. To find out the output g gives, you plug the input into the x slot. For example, if I input $x = 21$, the output—called $g(21)$—is found as follows:

$$g(21) = \frac{1}{3}(21) - 3$$
$$g(21) = 7 - 3 = 4$$

A *function* is a specific kind of relation. In a function, no input is allowed to give you more than one output. When one number goes in, only one matching number is allowed to come out. The relation g above is a function of x, because for every x you plug in, you can only get one result. If you plug in $x = 3$, you will always get –2. If you did it 50 times, you wouldn't suddenly get 101.7 as your answer on the forty-ninth try! Every input results in only one corresponding output. Different inputs can result in different outputs, for example, $g(3) \neq g(6)$. That's okay. You just can't get different answers when you plug in the same initial quantity.

The word *domain* is usually used to describe the set of inputs for a function. Any number that a function accepts as an appropriate input is part of the domain. For example, in the function s:{(–1,5),(1,6),(2,4)}, the domain is {–1,1,2}. The set of outputs to a function is called the *range*. The range of s is {4,5,6}.

Enough math for a second—let's relate this to real life. A person's height is a function of time. If I ask, "How tall were you at exactly noon today?" you could give only one answer. You couldn't respond "5 feet 6 inches" *and* "6 feet 1 inch," unless, of course, you lied on your driver's license.

Sometimes you'll plug more than a number into a function—you can also plug a *function* into another function. This is called composition of functions, and is not difficult to do. Simply start by evaluating the inner function and work your way out.

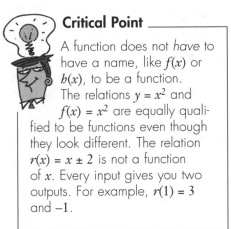

Critical Point

A function does not *have* to have a name, like $f(x)$ or $h(x)$, to be a function. The relations $y = x^2$ and $f(x) = x^2$ are equally qualified to be functions even though they look different. The relation $r(x) = x \pm 2$ is not a function of x. Every input gives you two outputs. For example, $r(1) = 3$ and –1.

Example 1: If $f(x) = \sqrt{x}$ and $g(x) = x + 6$, evaluate $g(f(25))$.

Solution: In this case, 25 is being plugged into f, and that output is in turn plugged into g. Start in the belly of the beast and evaluate $f(25)$. This is easy: $f(25) = \sqrt{25} = 5$. Now, plug this result into g:

$$g(5) = 5 + 6 = 11$$

Therefore, $g(f(25)) = 11$.

You've Got Problems

Problem 1: If $f(x) = \frac{x-1}{6}$, $g(x) = x^2 + 15$, and $h(x) = \sqrt[3]{x}$, evaluate $h(g(f(43)))$.

Sometimes in calculus, you run across a weird entity: the piecewise-defined function. This function is similar to Frankenstein's monster because it is created by sewing other functions together. Take a look at this messy thing:

$$f(x) = \begin{cases} 2x + 3, & x < 2 \\ x - 4, & x \geq 2 \end{cases}$$

Don't get confused—it's not that hard to understand. Basically, if you want to plug a number less than 2 into f, you plug it into $2x + 3$. For example, $f(1) = 2(1) + 3 = 5$. However, if your input is greater than or equal to 2, you plug it into $x - 4$ to get the output. For example, $f(10) = 10 - 4 = 6$. People sometimes get confused about those restrictions ($x < 2$ and $x \geq 2$). Remember that these restrictions are on the input (x) and not the output, and you'll be fine.

Talk the Talk

The **vertical line test** tells you whether or not a graph is a function. If any vertical line can be drawn through the graph that intersects that graph more than once, then the graph in question cannot be a function.

The last important thing you should know about functions is the *vertical line test*. This test is a way to tell whether or not a given graph is the graph of a function or not. All you have to do is draw imaginary vertical lines through the graph and note the number of times these lines hit the graph (see Figure 3.1). If any imaginary line can be drawn through the graph that hits it more than once, the graph cannot be a function.

Figure 3.1

No vertical line intersects the graph on the left more than once, so it is a function. However, some vertical lines hit the right-hand graph more than once, so it cannot be a function.

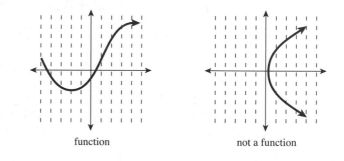

function   not a function

Functional Symmetry

Now that you know a thing or two about functions, you should also know some of the key classifications and buzzwords. If you throw these words around at parties, you'll surely impress your peers. Just think about how impressed they'd be with an offhand comment

Talk the Talk

A **symmetric** function looks like a mirror image of itself, typically across the x-axis, y-axis, or about the origin.

like, "That painting really exploits y-symmetry to show us our miniscule place in the world." Maybe you and I don't go to the same sorts of parties ….

A function has *symmetry* if it mirrors itself with respect to a fixed part of the coordinate plane. That sounds like a complicated concept, but it is not. Consider, for example, the graph of $y = x^2$.

Notice that the graph looks exactly the same on either side of the y-axis. This function is said to be y-symmetric, or even. There is an easy arithmetic test for y-symmetry that doesn't require the graph to complete.

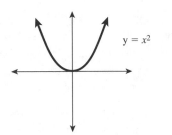

Figure 3.2

Feast your eyes on a graph that is symmetric about the y-axis.

Example 2: Determine whether or not $y = x^4 - 2x^2 + 1$ is *y*-symmetric.

Solution: Replace each of the *x*'s with $(-x)$ and simplify the equation:

$$y = \left(-x\right)^4 - 2\left(-x\right)^2 + 1$$
$$y = x^4 - 2x^2 + 1$$

Whenever a negative number is raised to an even power, the negative sign will be eliminated. Notice that our simplified result is the same as the original equation. When this happens, you know that the equation is, indeed, *y*-symmetric. In case you'd also like visual proof, here's the graph in Figure 3.3:

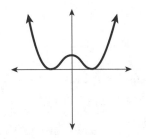

Figure 3.3

The graph of $y = x^4 - 2x^2 + 1$.

The other two major kinds of symmetry are *x*-symmetry and origin-symmetry, illustrated by the graphs in Figure 3.4.

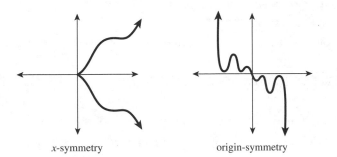

x-symmetry   origin-symmetry

Figure 3.4

Two other types of symmetry you may see. Note that most x-symmetric equations are not functions, because they fail the Vertical Line Test.

Very similar to y-symmetry, x-symmetry requires that the graph be identical above and below the x-axis. The test for x-symmetry is similar to y-symmetry, except that you plug in $(-y)$ for the y's in the equation instead of $(-x)$ for the x's. Again, if the equation reverts back to its original form when simplifying is over, then the equation is x-symmetric. If even one sign is different, then the equation is not x-symmetric.

Origin-symmetry is achieved when the graph does exactly the opposite thing on either side of the origin. In the above graph, notice that the origin-symmetric curve snakes down and to the right as x gets more positive, and up and to the left as x gets more negative. In fact, every turn in the second quadrant is matched and inverted in the fourth quadrant. To test an equation for origin-symmetry, replace all x's with $(-x)$ and all y's with $(-y)$. Once again, if the simplified equation matches your original equation, then that function is origin-symmetric. By the way, if a function is origin-symmetric, you can also classify it as an odd function.

> **You've Got Problems**
>
> Problem 2: Determine what kind of symmetry, if any, is evident in the graph of $y = \frac{x^3}{|x|}$.

Graphs to Know by Heart

During your study of calculus, you'll see certain graphs over and over again. Because of this, it's important to know them intuitively. You're already familiar with these functions, but make sure you know their graphs intimately, and it will save you time and frustration in the long run. Tell that someone special in your life that they can no longer possess your whole heart—they're going to have to share it with some math graphs. If they don't understand your needs, it wasn't meant to be for the two of you.

The descriptions are as follows:

◆ $y = x$: the most basic linear equation; has slope 1 and y-intercept 0; origin-symmetric; both domain and range are all real numbers

◆ $y = x^2$: the most basic quadratic equation; y-symmetric; domain is all real numbers; range is $y \geq 0$

◆ $y = x^3$: the most basic cubic equation; origin-symmetric; domain and range are all real numbers

◆ $y = |x|$: the absolute value function; returns the positive form of the input; y-symmetric; made of two line segments of slope -1 and 1, respectively; domain is all real numbers, range is $y \geq 0$

◆ $y = \sqrt{x}$: the square root function; has no symmetry; domain is $x \geq 0$ (you can't find the square root of numbers less than 0); range is $y \geq 0$

◆ $y = \frac{1}{x}$: no x- or y-intercepts; domain and range are both all real numbers except for 0

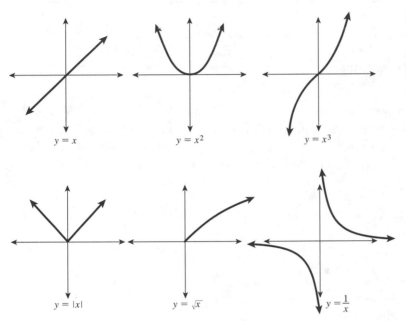

$y = x$ $y = x^2$ $y = x^3$

$y = |x|$ $y = \sqrt{x}$ $y = \frac{1}{x}$

Figure 3.5

The six most basic functions that will soon reside in your heart (specifically the left ventricle).

Constructing an Inverse Function

You have used inverse functions forever without even realizing it. They are the tools you break out to eliminate something unwanted in an equation. For example, how would you solve the equation $x^2 = 9$? To solve for x, you would take the square root of both sides to eliminate the squared term. This works because $y = \sqrt{x}$ and $y = x^2$ are inverse functions.

Mathematically speaking, f and g are inverse functions if …

$$f(g(x)) = g(f(x)) = x$$

In other words, plugging g into f and f into g leaves behind no trace of the function (not even forensic evidence), only x. Let's go back to $y = \sqrt{x}$ and $y = x^2$ for a second and show mathematically that they are inverse functions. If we plug these functions into each other, they will cancel out, leaving only x behind:

$$\left(\sqrt{x}\right)^2 = \sqrt{x^2} = x$$

Inverse functions have special notation. The inverse to a function $f(x)$ is written as $f^{-1}(x)$. This does *not* mean "f to the -1 power." It is read "the inverse of f" or "f inverse." I know the notation is a little confusing, since a negative exponent usually means that the indicated piece belongs in a different part of the fraction.

Now for some good news. It is very easy to create an inverse function. The word "easy" is usually very misleading when used by math teachers. In fact, whenever I qualified a class discussion with "Now, this is easy …," the students knew that it was going to be anything but. However, I wouldn't lie to you, would I? You decide as you read the next example.

Example 3: If $g(x) = \sqrt{2x + 5}$, find $g^{-1}(x)$.

Solution: For starters, replace the function notation $g(x)$ with y:

$$y = \sqrt{2x + 5}$$

Here's the key step: Reverse the x and y. In essence, this is what an inverse function does—it turns a function inside out so that the result has the spiffy property of canceling out the initial equation:

$$x = \sqrt{2y + 5}$$

Your goal now is to solve this equation for y, and you'll be done. In this problem, that means squaring both sides of the equation:

$$x^2 = 2y + 5$$

Now, subtract 5 from both sides and divide by 2 to finish solving for y:

$$x^2 - 5 = 2y$$
$$\frac{x^2 - 5}{2} = y$$

That is the inverse function. To finish, write it in proper inverse function notation:

$$g^{-1}(x) = \frac{x^2 - 5}{2}$$

Parametric Equations

With all this talk about functions, you might be leery of nonfunctions. Don't get all closed-minded on me. You can use something called *parametric equations* to express graphs, too, and they have the unique ability to represent nonfunctions (like circles) very easily. Parametric equations are pairs of equations, usually in the form of "$x =$" and "$y =$", that define points of the graph in terms of yet another variable, usually t.

What's a Parameter?

That definition's quite a mouthful, I know. To get a better understanding, let's look at an example of parametric equations:

$$x = t + 1$$
$$y = t - 2$$

Talk the Talk

Parametric equations define a graph in terms of a third variable, or parameter.

These two equations together produce one graph. To find that graph, you have to substitute a spectrum of things for the parameter t; each time you make a t substitution, you'll get a point on the graph. So a parameter is just a variable into which you plug numeric values to find coordinates on a parametric equation graph. For example, if you plug $t = 1$ into the equations, you get the following:

$$x = t + 1 = 1 + 1 = 2$$
$$y = t - 2 = 1 - 2 = -1$$

Therefore, the point $(2,-1)$ is on the graph. To get another point, I'll plug in $t = -2$, but you can actually plug in any real number for t:

$$x = -2 + 1 = -1$$
$$y = -2 - 2 = -4$$

A second point on the graph is $(-1,-4)$. You can see that this process takes a while. In fact, to get an exact graph, you'd probably have to plug in an infinite number of points.

Converting to Rectangular Form

Let's be honest, no one wants to plug in an infinite number of points. Even if you had the time to do that, you could definitely find something better to do. Therefore, it behooves us to learn how to translate from parametric form to the form we know and love, rectangular form. In the next example, we'll translate that set of parametric equations into something more manageable.

Example 4: Translate the parametric equations $x = t + 1$, $y = t - 2$ into rectangular form.

Solution: Begin by solving one of the equations for t. They're both pretty basic, so it doesn't matter which you choose. I'll pick the x equation so my result is in the form "$y = $". That makes it easier to graph:

$$x = t + 1$$
$$t = x - 1$$

Now we have t in terms of x. Therefore, you can replace the t in the y equation with $(x - 1)$, since we know that $t = x - 1$:

$$y = t - 2$$
$$y = (x - 1) - 2$$
$$y = x - 3$$

This is just a line in slope-intercept form, so our parametric equations' graph is just the line with slope 1 and y-intercept -3.

Figure 3.6

The graph of y = x – 3. *Note that the points (2,–1) and (–1,–4) are both on the graph, as we suspected from our work preceding Example 4.*

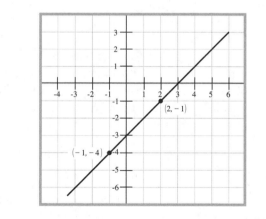

Problem 5: Put the parametric equations $x = t + 1$, $y = t^2 - t + 1$ into rectangular form.

The Least You Need to Know

- A relation becomes a function when each of its inputs can only result in one matching output.

- The inputs of a function comprise the domain and the outputs make the range.

- When a function is plugged into its inverse function (and vice versa), they cancel each other out.

- Parametric equations are defined "x =" and "y =" equations that contain a parameter, usually t.

4

Trigonometry: Last Stop Before Calculus

In This Chapter

◆ Characteristics of periodic functions

◆ The six trigonometric functions

◆ The importance of the unit circle

◆ Key trigonometric formulas and identities

Trigonometry, the study of triangles, has been around for a long time, creeping mysteriously from shadow to shadow and occasionally snatching unwary students into its razor-sharp clutches and causing the end of their mathematics careers. Few things cause people to panic like trig does, with the exception of TV weatherman Al Roker.

It is a commonly held belief that children on All Hallow's Eve historically have marched from door to door, sacks in hand, chiming, "Trig or Treat!" In response, homeowners would reward them with small protractors and compasses to avoid the wrath of neighborhood pranksters. If all the children went away happily, it was a good "sine" for harvest. However, this myth is definitely untrue, and I have gone off on a tangent.

Getting Repetitive: Periodic Functions

There are six major trigonometric functions, at least three of which you have probably heard: sine, cosine, and tangent. All of the trigonometric functions (even those that are offended because you haven't heard of them) are periodic functions. A *periodic function* has the unique characteristic that it repeats itself after some fixed period of time. Think of the rising of the sun as a periodic function—every 24 hours (a fixed amount of time) the sun appears on the horizon.

Talk the Talk —————

A periodic function's values repeat over and over, at the same rate and at the same intervals in time. The length of the horizontal interval after which the function repeats is called the **period**.

The amount of horizontal space it takes until the function repeats itself is called the *period*. For the most basic trigonometric functions (sine and cosine), the period is 2π. Look at the graph in Figure 4.1 of one period of $y = \sin x$.

Figure 4.1

One period of y = *sinx.*

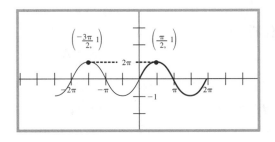

The graph of the sine function is a wave, reaching a maximum height of 1 and a minimum height of -1. On the piece of the graph shown above, the maximum height is reached at $x = -\frac{3\pi}{2}$ and $x = \frac{\pi}{2}$. The distance between these two points, where the graph repeats its value, is 2π. If that doesn't help you understand what is meant by period, consider the darkened portion of the graph.

This piece begins at the origin (0,0) and wiggles up and down, returning to a height of 0 when $x = 2\pi$. True, the graph hits a height of 0, repeating its value, when $x = \pi$, but it hasn't completed its period yet—that is only finished at $x = 2\pi$.

Talk the Talk —————

Coterminal angles have the same function value, because the space between them is a multiple of the function's period.

If you were to extend the graph of the sine function infinitely right and left, it would continue infinitely, redrawing itself every 2π. Because of this property of periodic functions, you can list an infinite number of inputs that have identical sine values. These are called *coterminal angles*, and the next example focuses on them.

Example 1: List two additional angles (one positive and one negative) that have the same sine value as $\frac{\pi}{4}$.

Solution: We know that sine repeats itself every 2π, so exactly 2π further up and down the x-axis from $\frac{\pi}{4}$, the value will be the same. To find these values, simply add 2π to $\frac{\pi}{4}$ in order to get one and subtract 2π from $\frac{\pi}{4}$ to get the other. In order to add and subtract the values, you'll have to get common denominators:

$$\frac{\pi}{4} + 2\pi = \frac{\pi}{4} + \frac{8\pi}{4} = \frac{9\pi}{4}$$

$$\frac{\pi}{4} - 2\pi = \frac{\pi}{4} - \frac{8\pi}{4} = -\frac{7\pi}{4}$$

Therefore, the angles $\frac{9\pi}{4}$ and $-\frac{7\pi}{4}$ are coterminal to $\frac{\pi}{4}$ and $\sin\frac{9\pi}{4} = \sin\left(-\frac{7\pi}{4}\right) = \sin\frac{\pi}{4}$.

Introducing the Trigonometric Functions

Time to meet the cast. There are six players in the drama we call trigonometry. You'll see a graph of each and learn a little something about the function. Whereas it's not extremely important to memorize the graphs of these functions, it's good to see how the graphs illustrate the functions' properties. So, here they are, in roughly the order of importance to you in your quest for calculus.

Sine (Written as $y = \sin x$)

The sine function is defined for all real numbers, and this unrestricted domain makes the function very trustworthy and versatile (see Figure 4.2). The range is $-1 \le y \le 1$, so all sine values fall within those boundaries. Notice that the sine function has a value of 0 whenever the input is a multiple of π. Sometimes, people get confused when memorizing unit circle values (more on the unit circle later in this chapter). If you remember the graph of sine, you can easily remember that $\sin 0 = \sin\pi = \sin 2\pi = 0$, because that's where the graph crosses the x-axis. The period of the sine function is 2π.

Cosine (Written as $y = \cos x$)

Cosine is the "cofunction" of sine (see Figure 4.3). (In other words, their names are the same, except one has a "co-" prefix, but I bet you figured that out.) As such, it looks very similar, possessing the same domain, range, and period. In fact, if you shift the entire graph of $y = \cos x$ a total of $\frac{\pi}{2}$ radians to the right, you get the graph of $y = \sin x$! The cosine has a value of 0 at all the "half-π's," such as $\frac{\pi}{2}$ and $\frac{3\pi}{2}$.

Figure 4.2

The sine function.

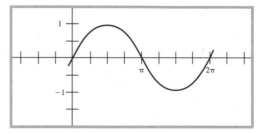

Critical Point _____

Throughout this book, I will refer to and evaluate trigonometric values in terms of radians, as they are used far more prevalently than degrees in calculus. Both degrees and radians are simply alternate ways to measure angles, just as Celsius and Fahrenheit are alternate ways to measure temperature. To get a rough idea of the conversion, remember that π radians = 180 degrees. If you want to convert from radians to degrees, multiply by $\frac{180}{\pi}$. For example, $\frac{\pi}{2}$ is equivalent to $\frac{\pi}{2} \cdot \frac{180}{\pi} = 90$ degrees. To convert from degrees to radians, multiply by $\frac{\pi}{180}$.

Tangent (Written as $y = \tan x$)

The tangent is defined as the quotient of the previous two functions: $\tan x = \frac{\sin x}{\cos x}$. Thus, to evaluate $\tan\frac{\pi}{4}$, you'd actually evaluate $\frac{\sin\frac{\pi}{4}}{\cos\frac{\pi}{4}}$ (which will equal 1 for those of you who are curious, but more about that later). Because the cosine appears in the denominator, the tangent will be undefined whenever the cosine equals 0, which (according to the last section) is at the half-π's (see Figure 4.4). Notice that the graph of the tangent has vertical *asymptotes* at these values. The tangent equals 0 at each midpoint between the asymptotes. The domain of the tangent excludes the "half-π's," $\left\{..., -\frac{3\pi}{2}, -\frac{\pi}{2}, \frac{\pi}{2}, \frac{3\pi}{2}, ...\right\}$, but the range is all real numbers. The period of the tangent is π—notice that graph completes one complete version of itself between $-\frac{\pi}{2}$ and $\frac{\pi}{2}$.

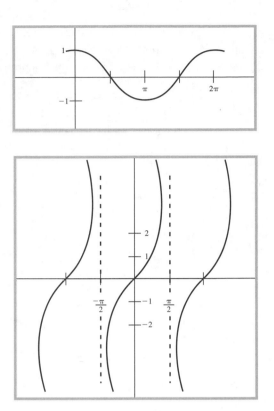

Figure 4.3
The cosine function.

Figure 4.4
The tangent function.

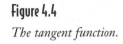

Talk the Talk

An **asymptote** is a line representing an unattainable value that shapes a graph. Because the graph cannot achieve the value, the graph typically bends toward that line forever and ever, yearning, reaching, stretching, but unable to reach it. A vertical asymptote typically indicates the presence of 0 in the denominator of a fraction. For example, the vertical line $x = \frac{\pi}{2}$ is a vertical asymptote of $y = \tan x$ because the tangent has 0 in the denominator whenever $x = \frac{\pi}{2}$.

Cotangent (Written as $y = \cot x$)

The cofunction of tangent, cotangent, is the spitting image of tangent, with a few exceptions (see Figure 4.5). It, too, is defined by a quotient: $\cot x = \frac{\cos x}{\sin x}$. In fact, the cotangent is technically the *reciprocal* of the tangent, so you can also write $\cot x = \frac{1}{\tan x}$. Therefore, this function is undefined whenever sin $x = 0$, which occurs at all the multiples of π: $\{\ldots, -2\pi, -\pi, 0, \pi, 2\pi, \ldots\}$, so the domain includes all real numbers except that set. The range, like that of the tangent, is all real numbers, and the period, π, also matches the tangent's.

Figure 4.5

The cotangent function.

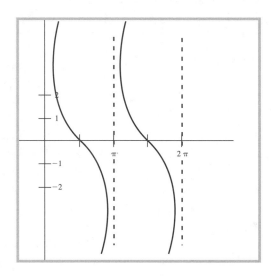

Secant (Written as $y = \sec x$)

The secant function is simply the reciprocal of the cosine, so $\sec x = \frac{1}{\cos x}$. Therefore, the graph of the secant is undefined (has vertical asymptotes) at the same places (and for the same reasons) as the tangent, since they both have the same denominator (see Figure 4.6). Hence, the two functions also have the same domain. Notice that the secant has no x-intercepts. In fact, it doesn't even come close to the x-axis, only venturing as far in as 1 and -1. That's a fascinating comparison: The cosine has a range of $-1 \le y \le 1$, but the secant has a range of $y \le -1$ together with $y \ge 1$—almost the exact opposite. Because the secant is based directly on the cosine, the functions have the same period, 2π.

Talk the Talk

The **reciprocal** of a fraction is the fraction flipped upside down (for example, the reciprocal of $\frac{7}{4}$ is $\frac{4}{7}$). The word *refliprocal* helps me remember what it means.

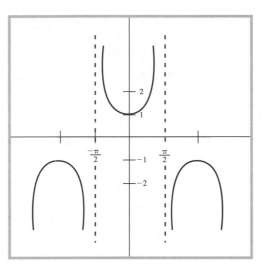

Figure 4.6
The secant function.

Critical Point _____

It's important to know how 0 affects a fraction. If 0 appears in the denominator of a fraction, that fraction is deemed "undefined." It is against math law to divide by 0, and the penalties are stiff (the same as ripping tags off of mattresses). On the other hand, if 0 appears in the numerator of a fraction (and not the denominator, too … that's still not allowed), then the fraction's value is 0, no matter how big a number is in the denominator.

Cosecant (Written as $y = \csc x$)

Very similar to its cofunction sister, this function has the same range and period as the secant, differing only in its domain. Because the cosecant is defined as the reciprocal of the sine, $\csc x = \frac{1}{\sin x}$, the cosecant will have the same domain as the cotangent, as they share the same denominator (see Figure 4.7).

In essence, four of the trig functions are based on the other two (sine and cosine), so those two alone are sufficient to generate values for the rest.

Example 2: If $\cos\theta = \frac{1}{3}$ and $\sin\theta = -\frac{\sqrt{8}}{3}$, evaluate $\tan\theta$ and $\sec\theta$.

CAUTION **Kelley's Cautions** _____

The cosecant is not the reciprocal of the cosine. Many times, people pair these because they have the same initial sound, but it's wrong. Similarly, the secant is not the reciprocal of the sine, even though they have the same initial sound.

Solution: Let's tackle these one at a time. First of all, you know that $\tan\theta = \frac{\sin\theta}{\cos\theta}$, so:

$$\tan\theta = \frac{-\dfrac{\sqrt{8}}{3}}{\dfrac{1}{3}}$$

Multiply the top and bottom by 3 to simplify the fraction and your final answer is ...

$$\tan\theta = \frac{-\dfrac{\sqrt{8}}{3}}{\dfrac{1}{3}} \cdot \frac{\dfrac{3}{1}}{\dfrac{3}{1}} = -\sqrt{8}$$

Now, on to secθ—this is even easier. Because you know that $\cos\theta = \frac{1}{3}$, and $\sec\theta = \frac{1}{\cos\theta}$ (because the secant is the reciprocal of the cosine):

$$\sec\theta = \frac{1}{\dfrac{1}{3}} = 3$$

Figure 4.7

The cosecant function.

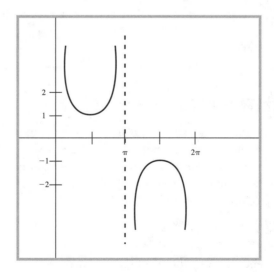

What's Your Sine: The Unit Circle

No one expects you to be able to evaluate most trigonometric expressions off the top of your head. If someone held a gun to my head and asked me to evaluate $\cos\frac{3\pi}{7}$ with an accuracy of .001, I would respond by calmly lying on the ground and beginning to draw a chalk outline around myself and preparing for death. I'd have no chance without a calculator or a Rainman-like ability for calculation. Most calculus classes, however, will require you to know certain trigonometric values without even a second thought.

These values are derived from something called the *unit circle*, a circle with a radius of length 1 that generates common cosine and sine values. You don't really have to know how to get those values (or how the unit circle works), but you should have these values memorized. Make flash cards, recite them with a partner, get a tattoo—whatever method you use to remember things—but memorize the unit circle values in the chart in Figure 4.8.

Talk the Talk

The **unit circle** is a circle whose radius is 1 unit that can be used to generate the most common values of sine and cosine. Rather than generating them each time you need them, it's best to simply memorize those common values.

Angle (radians)	cosine	sine	Angle (radians)	cosine	sine
0	1	0	π	-1	0
$\dfrac{\pi}{6}$	$\dfrac{\sqrt{3}}{2}$	$\dfrac{1}{2}$	$\dfrac{7\pi}{6}$	$-\dfrac{\sqrt{3}}{2}$	$\dfrac{-1}{2}$
$\dfrac{\pi}{4}$	$\dfrac{\sqrt{2}}{2}$	$\dfrac{\sqrt{2}}{2}$	$\dfrac{5\pi}{4}$	$-\dfrac{\sqrt{2}}{2}$	$-\dfrac{\sqrt{2}}{2}$
$\dfrac{\pi}{3}$	$\dfrac{1}{2}$	$\dfrac{\sqrt{3}}{2}$	$\dfrac{4\pi}{3}$	$-\dfrac{1}{2}$	$-\dfrac{\sqrt{3}}{2}$
$\dfrac{\pi}{2}$	0	1	$\dfrac{3\pi}{2}$	0	-1
$\dfrac{2\pi}{3}$	$-\dfrac{1}{2}$	$\dfrac{\sqrt{3}}{2}$	$\dfrac{5\pi}{3}$	$\dfrac{1}{2}$	$-\dfrac{\sqrt{3}}{2}$
$\dfrac{3\pi}{4}$	$-\dfrac{\sqrt{2}}{2}$	$\dfrac{\sqrt{2}}{2}$	$\dfrac{7\pi}{4}$	$\dfrac{\sqrt{2}}{2}$	$-\dfrac{\sqrt{2}}{2}$
$\dfrac{5\pi}{6}$	$-\dfrac{\sqrt{3}}{2}$	$\dfrac{1}{2}$	$\dfrac{11\pi}{6}$	$\dfrac{\sqrt{3}}{2}$	$\dfrac{-1}{2}$

Figure 4.8

The confounded unit circle, a necessary evil to calculus. Memorize it now and avoid trauma in the future.

If you're having trouble remembering the unit circle, look for patterns. If you absolutely refuse to memorize these values and it's okay with your instructor that you don't, at the very least keep this chart in a handy place, because you'll find yourself consulting it often.

Now that you know the unit circle and all kinds of crazy stuff about trig functions, your powers have increased. (Just make sure you always use them for good, not evil.) In fact, you are able to evaluate a lot more functions, as demonstrated by the next example.

Example 3: Find the value of $\cos\dfrac{23\pi}{4}$.

Solution: We only know the values of sine and cosine from 0 radians to $\frac{11\pi}{6}$ radians. Clearly, $\frac{23\pi}{4}$ is much too large to fit in this limited interval. However, because cosine is a periodic function, its values will repeat. Since cosine's period is 2π, you can find a coterminal angle to $\frac{23\pi}{4}$ which does appear in our unit circle chart and evaluate that one instead—the answer will be the same.

According to Example 1 in this chapter, all you have to do is add or subtract the period (again, it is 2π for cosine) and you'll get a coterminal angle. I am looking for a smaller angle than $\frac{23\pi}{4}$, so I'll subtract 2π. Don't forget to get common denominators to subtract correctly:

$$\frac{23\pi}{4} - 2\pi = \frac{23\pi}{4} - \frac{8\pi}{4} = \frac{15\pi}{4}$$

That's still too big (the largest $\frac{\pi}{4}$ angle I have memorized is $\frac{7\pi}{4}$), so I have to subtract again:

$$\frac{15\pi}{4} - \frac{8\pi}{4} = \frac{7\pi}{4}$$

Because $\frac{7\pi}{4}$ and $\frac{23\pi}{4}$ are coterminal, $\cos\frac{7\pi}{4} = \cos\frac{23\pi}{4}$, which we know to be $\frac{\sqrt{2}}{2}$.

You've Got Problems

Problem 1: Evaluate $\cos\frac{14\pi}{4}$ using a coterminal angle and the unit circle.

You might be interested in how the unit circle originates and how the previous values are derived. Here's a quick explanation. A unit circle is just a circle with radius one, and we'll center it at the origin. Now, draw a ray from the origin that makes a 30-degree $\left(\frac{\pi}{6}\text{-radian}\right)$ angle with the positive x-axis in the first quadrant, and mark the point where the ray intersects the circle. The coordinates of that point are, respectively, the cosine and sine of $\frac{\pi}{6}$. To find the coordinates of the point, find the lengths of the legs of the right triangle shown in Figure 4.9.

Incredibly Important Identities

An identity is an equation that is always true, regardless of the input. It's easy to tell that, according to this definition, $x + 1 = 7$ is not an identity, because it is only true when $x = 6$. However, consider the equation $2(x - 1) + 3 = 2x + 1$. If you plug in $x = 0$, you get $1 = 1$, which is definitely true, wouldn't you agree? Try plugging any real number, and you'll get another true statement. Thus, $2(x - 1) + 3 = 2x + 1$ is an identity.

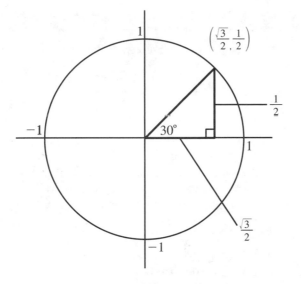

Figure 4.9

This right triangle, inscribed on the unit circle, determines the values of cosine and sine for the angle at the origin.

It is worth mentioning that it is a very stupid identity. You are not going to impress anyone by showing that equation off. With only two seconds' worth of work, you can simplify the left side and show the two sides equal. Most math identities are much more useful because it's not immediately obvious that they are true. Specifically, trigonometric identities help you rewrite equations, simplify expressions, and justify answers to equations. With this in mind, we'll explore the most common trigonometric identities. They're worth memorizing if you don't know them already.

Pythagorean Identities

The three most important of all the trig identities are the Pythagorean identities. They are named as such because they are created with the Pythagorean theorem. Remember that little nugget from geometry? It said that the sum of the squares of the legs of a right triangle is equal to the square of the hypotenuse: $a^2 + b^2 = c^2$. Ever since I taught these three, I have named them the Mama, Papa, and Baby theorems respectively (after the Three Bears story), both for entertainment purposes and because they have no commonly accepted names. If I were in the company of math nerds, however, I wouldn't use the terms if I were you. They will sneer and scowl at you.

- ◆ Mama theorem: $\cos^2 x + \sin^2 x = 1$
- ◆ Papa theorem: $1 + \tan^2 x = \sec^2 x$
- ◆ Baby theorem: $1 + \cot^2 x = \csc^2 x$

You've Got Problems

Problem 2: The Mama theorem is an identity, and therefore true for every input. Show that it is true for $x = \frac{2\pi}{3}$.

Critical Point

Those *Wizard of Oz* fans out there may remember that the Scarecrow spouts a formula when the Great and Powerful Oz grants him a brain. He states, "The sum of the square roots of any two sides of an isosceles triangle is equal to the square root of the remaining side." This is a false statement! He was probably supposed to quote the Pythagorean theorem, since it is one of the most recognizable theorems to the general public, but missed it quite badly. Perhaps Oz was not so powerful after all. The Tin Man's string of failed marriages and the Cowardly Lion's lack of success as a motivational speaker offer further evidence to Oz's lackluster gift giving.

Now, let's see how trigonometric functions and identities can make our lives easier. With a little knowledge of trigonometry and a little bit of elbow grease, even ugly expressions can be made beautiful.

Example 4: Simplify the trigonometric expression $\frac{\cos^2 x}{\sin x} + \sin x$ using a Pythagorean identity.

Solution: One of these terms is a fraction. You know that you must first have common denominators in order to add fractions, so multiply the second term by $\frac{\sin x}{\sin x}$ to get his-and-hers matching denominators of $\sin x$:

$$\frac{\cos^2 x}{\sin x} + \frac{\sin x}{1} \cdot \frac{\sin x}{\sin x}$$

$$\frac{\cos^2 x}{\sin x} + \frac{\sin^2 x}{\sin x}$$

Now that the denominators match, we can perform the addition in the numerator while leaving the denominator alone:

$$\frac{\cos^2 x + \sin^2 x}{\sin x}$$

That doesn't look any easier! Hold on a second. The numerator looks just like the Mama theorem, and according to the Mama theorem we know that $\cos^2 x + \sin^2 x = 1$. Therefore, substitute 1 in for the numerator:

$$\frac{1}{\sin x}$$

We could stop there, but we're on a roll! We also know that $\frac{1}{\sin x} = \csc x$, since the cosecant is the reciprocal of the sine. Therefore, the final answer is $\csc x$.

Kelley's Cautions

The notation $\cos^2 x$ is shorthand notation for $(\cos x)^2$. It wouldn't make any sense for the letters cos to be squared, as you might expect. The shorthand notation is used to avoid having to write those extra parentheses.

Double-Angle Formulas

These identities allow you to write trigonometric expressions containing double angles (such as $\sin 2x$ and $\cos 2\theta$) into equivalent single-angle expressions. In other words, these expressions eliminate the x coefficient of 2 inside a trigonometric expression.

- $\sin 2x = 2\sin x \cos x$ (This is the simplest double-angle formula, and memorizing it is a snap.)

- $\cos 2x = \cos^2 x - \sin^2 x$
  $$= 2\cos^2 x - 1$$
  $$= 1 - 2\sin^2 x$$

The cosine double-angle is a little trickier—there are actually three different things which can be substituted for $\cos 2x$. You should choose which to substitute in based on the rest of the problem. If there seem to be a lot of sines in the equation or expression, use the last of the three, for example.

There isn't a whole lot to understand about double-angle formulas. You should just be ready to recognize them at a moment's notice, as problems very rarely contain the warning label, "Caution: This problem will require you to know basic trig double-angle formulas. Keep away from eyes. May pose a choking hazard to children under 3." Watch how slyly these suckers slip in there.

Example 5: Factor and simplify the expression $\cos^4\theta - \sin^4\theta$.

Solution: This expression is the difference of perfect squares, so it can be factored as follows: $(\cos^2\theta + \sin^2\theta)(\cos^2\theta - \sin^2\theta)$. Notice that the left-hand quantity is equal to 1,

Critical Point

There are a lot of trig identities—not just the few in this chapter—but you'll use these far more than all the rest put together.

according to the Mama theorem, and the right-hand quantity is equal to cos2x, according to our double-angle formulas. Therefore, we can substitute those values to get $(1)(\cos 2x) = \cos 2x$.

You've Got Problems

Problem 3: Factor and simplify the expression $2\sin x\cos x - 4\sin^3 x \cos x$.

Solving Trigonometric Equations

The last really important trig skill you need to possess is the ability to solve trigonometric equations. A word of warning: Some math teachers get very bent out of shape when discussing trig equations. You will have to read the directions to these sorts of problems very carefully to make sure to answer the exact question being asked of you.

Critical Point

When intervals are specified as $[0,2\pi)$, that is shorthand for $0 \le x < 2\pi$. The two numbers in the notation represent the lower and higher boundaries of the acceptable interval, and the bracket or parenthesis tells you if that boundary is included in the interval or not. If it's a bracket, that boundary is included, but not so with a parenthesis. In interval notation, the expression $x \ge 7$ looks like $[7,\infty]$. Because there is no upper bound, you write infinity. If infinity is one of the boundaries, you always use a parenthesis next to it.

Kelley's Cautions

If your instructor demands one answer per equation, eliminate all of your solutions except for the one (and there will only be one) that falls into the appropriate range. That range is $-\frac{\pi}{2} \le \theta \le \frac{\pi}{2}$ for sine, tangent, and cosecant; for cosine, cotangent, and secant, use the interval of $0 \le \theta \le \pi$.

Each of my examples will ask for the solution to the trigonometric equation on the interval $[0, 2\pi]$. Therefore, there may be multiple answers. Some instructors will demand that you write the specific correct answer for each equation. In other words, although there may be many angles that solve the problem, they only accept one or two answers. This answer falls within a specific range, and as long as you learn the appropriate range for each trigonometric function, you'll be okay. The best approach is to ask if they'll require answers on a certain interval or if they expect only the answer on the appropriate range.

The procedure for solving trigonometric equations is not unlike solving regular equations. However, the final step often requires you to remember the unit circle!

Example 6: Solve the equation $\cos2x - \cos x = 0$ on the interval $[0, 2\pi]$.

Solution: First of all, we want to eliminate the double-angle formula so that all of the terms are single-angles. Because we are replacing $\cos2x$, there are three options, but I will choose the $2\cos^2x - 1$ option since the problem also contains another cosine term:

$$(2\cos^2x - 1) - \cos x = 0$$
$$2\cos^2x - \cos x - 1 = 0$$

Now you can factor this equation. (If you're having trouble, think of the equation as $2w^2 - w - 1$ and factor that, substituting in $w = \cos x$ when you're finished.)

$$(2\cos x + 1)(\cos x - 1) = 0$$

Like any other quadratic equation being solved using the factoring method, set each factor equal to 0 and solve:

$$2\cos x + 1 = 0 \text{ and } \cos x - 1 = 0$$
$$\cos x = -\frac{1}{2} \text{ and } \cos x = 1$$

To finish, ask yourself, "When is the cosine equal to $-\frac{1}{2}$ and when is it 1?" The question asks for all answers on $[0, 2\pi]$, so give all the correct answers on the unit circle:

$$x = \frac{2\pi}{3}, \frac{4\pi}{3}, \text{ and } 0$$

All three answers should be given.

If your instructor requires only answers on the appropriate ranges, your solutions would be $x = \frac{2\pi}{3}$ and $x = 0$. You'd throw out $x = \frac{4\pi}{3}$ because it does not fall in the correct cosine range of $0 \le \theta \le \pi$. In this case, it's okay to have a total of two answers, since each of the individual, smaller equations only has one answer.

You've Got Problems

Problem 4: Solve the equation $\sin2x + 2\sin x = 0$ and provide all solutions on the interval $[0, 2\pi]$.

The Least You Need to Know

◆ The six basic trigonometric functions are: sine, cosine, tangent, cotangent, secant, and cosecant.

◆ Sine and cosine's values are used when finding the values for the other four trig functions.

◆ There are some angles on the interval $[0, 2\pi]$ whose cosine and sine you should have memorized.

◆ Trigonometric identities help you simplify trig expressions and solve trig equations.

Part 2

Laying the Foundation for Calculus

I have some good news and some bad news. The good news is that we'll be dealing with relatively easy functions for the remainder of calculus. The bad news is that we need to mathematically define exactly what we mean by "easy." When math people sit down to define things, you know that theorems are going to start flying around, and calculus is no different.

When we say "easy" functions, we really mean continuous ones. In order to be continuous, the function can't contain holes and isn't allowed to have any breaks in it. That sounds nice, but math people like their definitions more specific (read: complicated) than that. In order to define "continuous," we'll first need to design something called a "limit." During this part of the book, you'll learn what a limit is, how to evaluate limits for functions, and how to apply that to design a definition for continuity.

Take It to the Limit

In This Chapter

◆ Understanding what a limit is

◆ Why limits are needed

◆ Approximating limits

◆ One-sided and general limits

When most people look back on calculus after completing it, they wonder why they had to learn limits at all. For some, it's like getting all of their teeth pulled just for the fun of it. After a brief limit discussion at the start of the course, there are very few times that limits return, and when they do, it is only for a brief cameo role in the topic at hand. However, limits are extremely important in the development of calculus and in all of the major calculus techniques, including differentiation, integration, and infinite series.

As I discussed in Chapter 1, limits were the key ingredient in the discovery of calculus. They allow you to do things that ordinary math gets cranky about. In practice, limits are many students' first encounter with a slightly philosophical math topic, answering questions like, "Even though this function is undefined at this *x*-value, what height did it *intend* to reach?" This chapter will give you a great intuitive feel of what a limit is and what it means for a function to have a limit; the next chapter will help you evaluate limits.

One final note: The official limit definition is called the delta-epsilon definition of limits. It is very complex, and is based on high-level mathematics. A discussion of this rigorous mathematical concept is not beneficial, so it is omitted here. In essence, it is possible to be a great driver without having to understand every principle of the combustion engine.

What Is a Limit?

When I first took calculus in high school, I was hip-deep in evaluating limits via tons of different techniques before I realized that I had no idea what exactly I was doing, or why. I am one of those people who needs some sort of universal understanding in a math class, some sort of framework to visualize why I am undertaking the process at hand. Unfortunately, calculus teachers are notorious for explaining *how* to complete a problem (outlining the steps and rules) but not explaining *what the problem means*. So, for your benefit and mine, we'll discuss what a limit actually is before we get too nutty with the math part of things.

Let's start with a simple function: $f(x) = 2x + 5$. We know that this is a line with slope 2 and y-intercept 5. If I plug $x = 3$ into the function, the output will be $f(3) = 2 \cdot 3 + 5 = 11$. Very simple, everyone understands, everyone's happy. What else does this mean, however? It means that the point (3,11) belongs to the relation and function I call f. Furthermore, it means that the point (3,11) falls on the graph of f, as evidenced in Figure 5.1.

Figure 5.1

The point (3,11) falls on the graph of f.

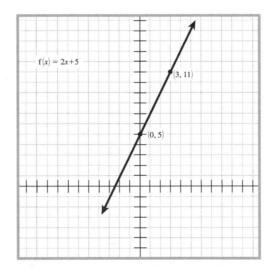

All of this seems pretty obvious, but let's change the way we talk just a little to prepare for limits. Notice that as we get closer and closer to $x = 3$, the height of the graph gets closer and closer to $y = 11$. In fact, if we plug $x = 2.9$ into f, we get $f(2.9) = 2(2.9) + 5 = 10.8$. If I plug in $x = 2.95$, the output is 10.9. Inputs close to 3 give us outputs close to 11, and the closer the input is to 3, the closer the output is to 11.

Even if we didn't know that $f(3) = 11$ (say for some reason we were forbidden by our evil step-godmother, as was Cinderella), we could still figure out what it would *probably* be by plugging in an insanely close number like 2.99999. I'll save you the grunt work and tell you that $f(2.99999) = 10.99998$. It's pretty obvious that f is headed straight for the point $(3,11)$, and that's what is meant by a limit.

A *limit* is the intended height of a function at a given value of x, whether or not the function actually reaches that height at the given x. In the case of f, we know that f does reach the value of 11 when $x = 3$, but that doesn't have to be the case for a limit to exist. Remember that a limit is the height a function *intends* to reach.

Talk the Talk

A **limit** is the height a function *intends* to reach at a given x value, whether or not it actually reaches it.

Can Something Be Nothing?

You may ask, "How am I supposed to know what a function *intends* to do? I don't even know what I intend to do." Luckily, functions are a little more predictable than people, but more on that later. For now, let's look at a slightly harder problem involving limits, but before we do, let's first discuss how a limit is written in calculus.

In our previous example, we determined that the limit, as x approaches 3, of $f(x)$ equals 11, because the function approached a height of 11 as we plugged in x values closer and closer to 3. As it seems with everything else, calculus has a shorthand notation for this:

$$\lim_{x \to 3} f(x) = 11$$

This is read, "The limit, as x approaches 3, of $f(x)$ equals 11." The tiny 3 is the number you're approaching, $f(x)$ is the function in question, and 11 is the intended height of f at 3. Now, let's look at a slightly more involved example.

Figure 5.2 is the graph of $g(x) = \frac{x^2 - x - 6}{x + 2}$. Clearly, the domain of g cannot contain $x = -2$, because that causes 0 in the denominator, and that is just plain yucky.

Notice that the graph of g has a hole at the evil value of $x = -2$, but that won't stop us. We're going to evaluate the limit there. Remember, the function doesn't actually have to exist at a certain

Critical Point

If you substitute $x = -2$ into $g(x) = \frac{x^2 - x - 6}{x + 2}$, you get $\frac{0}{0}$, which is said to be in "indeterminate form." Typically, a result of $\frac{0}{0}$ means that a hole appears in the graph at that value of x, which is the case with g.

point for a limit to exist—the function only has to have a clear height it intends to reach. Clearly, the function has an intended height it wishes to reach when $x = -2$ in the graph—there's a gaping hole at that exact spot, in fact.

Figure 5.2

The graph of

$$g(x) = \frac{x^2 - x - 6}{x + 2}.$$

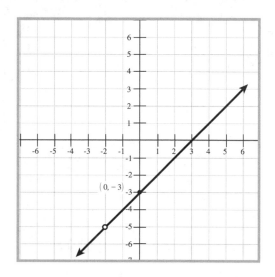

 Critical Point _____

We will evaluate limits like $\lim_{x \to 3} f(x)$ and $\lim_{x \to -2} g(x)$ in the next chapter without having to resort to the "plug in an insanely close number" technique. In this chapter, focus with me on the idea of a limit, and we'll get to the computational part soon enough.

How can we evaluate $\lim_{x \to -2} g(x)$? Just like we did in the previous example, we'll plug in a number insanely close to $x = -2$, in this case, $x = -1.99999$. Again, I'll do the grunt work for you (you can thank me later): $g(-1.99999) = -4.99999$. Even a knucklehead like me can see that this function intends to go to a height of -5 on the function g when $x = -2$.

Therefore, $\lim_{x \to -2} g(x) = -5$, even though the point $(-2,-5)$ does not appear on the graph of g. This is one example of a limit existing because a function intends to go to a height despite not actually reaching that height.

One-Sided Limits

Occasionally, a function will intend to reach two different heights at a given x, one height as you come from the left side, and one height as you come from the right side. We can still describe these one-sided intended heights, using *left-hand* and *right-hand limits*. To better understand this bizarre function behavior, look at the graph of $h(x)$ in Figure 5.3.

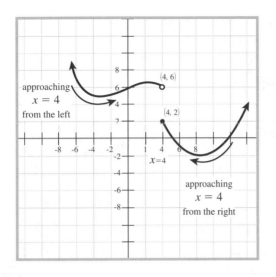

Figure 5.3

The graph of h(x) *consists of both pieces; a graph like this is usually the result of a piecewise-defined function.*

This graph does something very wacky at $x = 4$: It breaks. Trace your finger along the graph as it approaches $x = 4$ from the left. What height is your finger approaching as you get close to (but don't necessarily reach) $x = 4$? You are approaching a height of 6. This is called the left-hand limit and is written like this:

$$\lim_{x \to 4^-} h(x) = 6$$

The little negative sign in the exponent indicates that you should only be interested in the height the graph approaches as you travel along the graph from the left-hand side. If you trace your finger along the other portion of the graph, this time toward $x = 4$ from the right, you'll notice that you approach a height of 2 when you get close to $x = 4$. This is, as you may have guessed, the right-hand limit for $x = 4$, and it is written as follows:

$$\lim_{x \to 4^+} h(x) = 2$$

Talk the Talk

A **left-hand limit** is the height a function intends to reach as you approach the given x value *from* the left; the **right-hand limit** is the intended height as you approach *from* the right.

Critical Point

To keep from confusing right- and left-hand limits, remember the key word: *from*. A left-hand limit is the height toward which you're heading as you approach the given x-value *from* the left, not as you go *toward* the left on the graph.

Until now, we have only spoken of a general limit (in other words, a limit that doesn't involve a direction, like from the right or left). Most of the time in calculus, you will worry about general limits, but in order for general limits to exist, right- and left-hand limits must also be present; this we learn in the next section, which will tie together everything we've discussed so far about limits. Can you feel the electricity in the air?

When Does a Limit Exist?

If you don't understand anything else in this chapter, make sure to understand this section. It contains the two essential characteristics of limits: when they exist and when they don't exist. If you've understood everything so far, you're on the verge of understanding your first major calculus topic. I'm so proud of you … I remember when you were only *this* tall ….

Here's the key to limits: In order for a limit to exist on a function f at some x value (we'll give it a generic name like $x = c$), then three things must happen:

1. The left-hand limit must exist at $x = c$.

2. The right-hand limit must exist at $x = c$.

3. The left- and right-hand limits at c must be equal.

In calculus books, this is usually written like this: If $\lim_{x \to c^-} f(x) = \lim_{x \to c^+} f(x)$, then $\lim_{x \to c} f(x)$ exists and is equal to the one-sided limits.

The diagram in Figure 5.4 will help illustrate the point.

Figure 5.4

Yet another hideous graph called f(x). *Can you spot where the limit doesn't exist?*

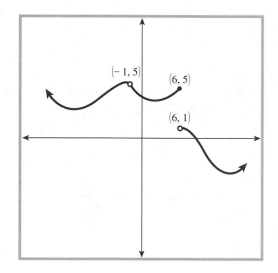

There are two interesting x-values on this graph: $x = -1$ and $x = 6$. At one of those values, a general limit exists, and at the other, no general limit exists. Can you figure out which is which using the above guidelines?

You're reading ahead aren't you? Well, stop it. Don't read any more until you've actually tried to answer the question I've asked you. Do it. I'm watching!

The answer: $\lim\limits_{x \to -1} f(x)$ exists and $\lim\limits_{x \to 6} f(x)$ does not. Remember, in order for a limit to exist, the left- and right-hand limits must exist at that point and be equal. As you approach $x = -1$ from the left and right sides, each time you are heading toward a height of 5, so the two one-sided limits exist and are equal, and we can conclude that $\lim\limits_{x \to -1} f(x) = 5$ (i.e., the general limit as x approaches -1 on $f(x)$ is equal to 5).

However, this is not the case when we approach $x = 6$ from the right and left. In fact, $\lim\limits_{x \to 6^-} f(x) = 5$, whereas $\lim\limits_{x \to 6^+} f(x) = 1$. Because those one-sided limits are unequal, we say that no general limit exists at $x = 6$, and that $\lim\limits_{x \to 6} f(x)$ does not exist.

Critical Point

Visually, a limit exists if the graph does not break at that point. For the graph $f(x)$ in question, a break occurs at $x = 6$ but not $x = -1$, which means a limit doesn't exist at the break but can exist at the hole in the graph. Remember that a limit can exist even if the function doesn't exist there—as long as the function intended to go to a specific height from each direction (as it did—it wanted to go to a height of 5), the limit exists.

When Does a Limit Not Exist?

Now, we need to talk about when limits do not exist. You already know one instance in which limits don't exist, so you're one-third of the way there.

◆ A general limit does not exist if the left- and right-hand limits aren't equal.

In other words, if there is a break in the graph of a function, and the two pieces of the function don't meet at an intended height, then no general limit exists there. In Figure 5.5, $\lim\limits_{x \to c} g(x)$ does not exist because the left- and right-hand limits are unequal.

◆ A general limit does not exist if a function increases or decreases infinitely at a given x-value (i.e., the function increases or decreases without bound).

In order for a general limit to exist, the function must approach some fixed numerical height. If a function increases or decreases infinitely, then no limit exists. In Figure 5.6, $\lim\limits_{x \to c} h(x)$ does not exist because $h(x)$ has a vertical asymptote at $x = c$, causing the function to increase without bound there.

Figure 5.5

The graph of g(x).

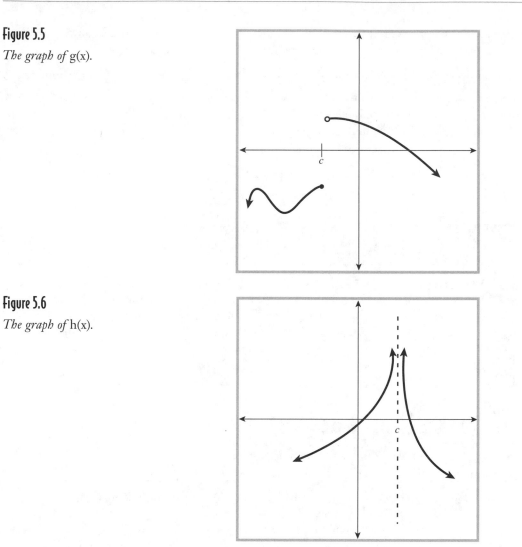

Figure 5.6

The graph of h(x).

◆ A general limit does not exist if a function oscillates infinitely, never approaching a single height.

This is rare, but sometimes a function will continually wiggle back and forth, never reaching a single numeric value. If this is the case, then no general limit exists. Because this is so rare, most calculus books give the same example when discussing this eventuality, and I will be no different (math peer pressure is harsh, let me tell you). No general limit exists at $x = 0$ in Figure 5.7 because the function never settles on any one value the closer you get to $x = 0$.

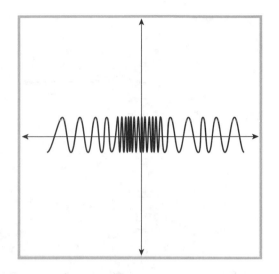

Figure 5.7

The graph of
$y = \sin\frac{1}{x}$; $\lim\limits_{x \to 0} \sin\frac{1}{x}$ *does not exist.*

Example: A certain function $f(x)$ is defined by the graph in Figure 5.8. Based on the graph and your amazing knowledge of limits, evaluate the limits that follow. If no limit exists, explain why.

a) $\lim\limits_{x \to -4^+} f(x)$

Solution: As you approach $x = -4$ from the right, the function increases without bound. You have two ways to write your answer; either say that the limit does not exist because the function increases infinitely, or write $\lim\limits_{x \to -4^+} f(x) = \infty$.

b) $\lim\limits_{x \to 4} f(x)$

Solution: As you approach $x = -4$ from the right and left, the function approaches a height of 3. Therefore, the general limit exists and is 3.

c) $\lim\limits_{x \to -3} f(x)$

Solution: No general limit exists here because the left-hand limit (1) does not equal the right-hand limit (–3).

Kelley's Cautions

If a graph has no general limit at one of its x-values, that does not affect any of the other x-values. For example, in the graph of $y = \sin\frac{1}{x}$, a general limit exists for every other x-value in its domain—only at $x = 0$ is there no general limit.

Critical Point

Giving a limit answer of ∞ or $-\infty$ is equivalent to saying that the limit does not exist. However, by answering with ∞ , you are also (1) explaining why the limit doesn't exist, and (2) specifically detailing whether the function increased or decreased infinitely there.

Figure 5.8

The hypnotic graph of f(x). *Mortal men may turn to stone upon encountering its terrible visage.*

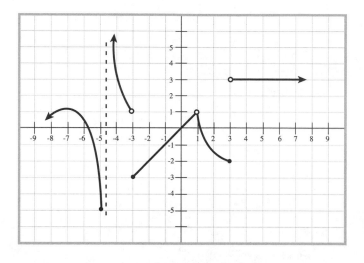

You've Got Problems

Here are a couple of limits to try on your own based on the graph of $f(x)$ used in the example in Figure 5.8.

1. $\lim\limits_{x \to -4^-} f(x)$

2. $\lim\limits_{x \to 3} f(x)$

3. $\lim\limits_{x \to 1} f(x)$

The Least You Need to Know

◆ The limit of a function at a given x-value is the height the function intends to reach there.

◆ A function can have a limit at an x-value even if the function has a hole there.

◆ A function cannot have a limit where its graph breaks.

◆ If a function's left- and right-hand limits exist and are equal for a certain $x = c$, then a general limit exists at c.

◆ A limit does not exist in the cases of infinite function growth or oscillation.

Chapter 6

Evaluating Limits Numerically

In This Chapter

◆ Three easy methods for finding limits

◆ Limits and asymptotes

◆ Finding limits at infinity

◆ Trig and exponential limit theorems

Now you know what a limit is, when a limit exists, and when it doesn't. However, the question of how to actually evaluate limits remains. In Chapter 5 we approximated limits by plugging in x values insanely close to the number we were approaching, but that got tedious quickly. As soon as you have to raise numbers like 2.999999 to various exponents, it becomes clear that we either need a better way or a giant bottle of aspirin.

Good news: There are lots of better ways, and this chapter will lead you through all of the major processes to evaluate limits and the important limit theorems you should memorize. For those of you who were uncomfortable with math turning a little conceptual and philosophical there for a little bit, don't worry—we're back to comfortable, familiar, soft, fuzzy, and predictable math techniques and formulas.

All of that theory you learned in the last chapter will resurface to some degree in Chapter 7, when we discuss continuity of functions, so keep it fresh in your mind. A lot of our discussion about limits will get hazy quickly when you move on to derivatives and integrals as the book progresses. Make sure you come back and review these early topics often throughout your calculus course to keep them fresh in your brain.

The Major Methods

The vast majority of limits can be evaluated by using one of three techniques: substitution, factoring, and the conjugate method. Usually, only one of these techniques will work on a given limit problem, so you should try one method at a time until you find one that works. Because I am efficient (understand, by that I mean extremely lazy) I always try the easiest method first, and only move on to more complicated methods if I absolutely have to. As such, I'll present the methods from easiest to hardest.

Substitution Method

Prepare yourself … you're going to weep with uncontrollable joy when I tell you this. Many limits can be evaluated simply by plugging the x value you're approaching into the function. The fancy term for this is the substitution method (or the direct substitution method).

Example 1: Evaluate $\lim_{x \to 4}\left(x^2 - x + 2\right)$.

Solution: In order to evaluate the limit, simply plug the number you're approaching (4) in for the variable:

$$4^2 - 4 + 2 = 16 - 2 = 14$$

 Critical Point

When I say that the methods of evaluating limits are listed from easiest to hardest, I should qualify it by saying that "hard" is not a good word choice; none of these methods is hard. The number of steps increases slightly from one method to the next, but these methods are easy.

According to the substitution method,

$$\lim_{x \to 4}\left(x^2 - x + 2\right) = 14$$

That was too easy! Just to make sure it actually worked, let's check the answer by looking at the graph of $y = x^2 - x + 2$ in Figure 6.1.

As we approach $x = 4$ from either the left or the right, the function clearly heads toward a height of 14, which, as we know, guarantees that the general limit exists and is 14. It worked! Huzzah!

If every limit problem in the world could be solved using substitution, there would probably be no need for Prozac. However (and there's always a however, isn't there?), sometimes substitution cannot be used. In such cases, you should resort to the next method of evaluating limits: factoring.

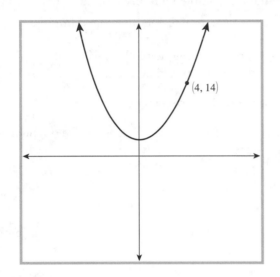

Figure 6.1

Use the graph of
y = x² – x + 2 to visually
verify the limit at x = 4.

You've Got Problems

Problem 1: Evaluate the following limits using substitution:

(a) $\lim\limits_{x \to \pi} \dfrac{\cos x}{x}$

(b) $\lim\limits_{x \to -2} \dfrac{x^2 + 1}{x^2 - 1}$

Factoring Method

Consider the function $f(x) = \dfrac{x^2 - 9}{x + 3}$. What if you want to find the limit of f as x approaches –3? Well, if you try to use substitution to find the limit, bad things happen:

$$\lim_{x \to -3} \frac{x^2 - 9}{x + 3} = \frac{(-3)^2 - 9}{-3 + 3}$$
$$= \frac{0}{0}$$

What kind of an answer is $\frac{0}{0}$? A gross one, that's for sure. Remember that we can't have 0 in the denominator of a fraction; that's not allowed. Clearly, then, the limit is not $\frac{0}{0}$, but that answer does tell us two things:

1. You must use a different method to find the limit, because …

2. … the function likely has a hole at the x value you substituted into the function.

The best alternative to substitution is the factoring method, which works just beautifully in this case. In the next example, we'll find this troubling limit.

Example 2: Evaluate $\lim\limits_{x \to -3} \frac{x^2 - 9}{x + 3}$ using the factoring method.

Solution: To begin the factoring method, factor! It makes sense, since the numerator is the difference of perfect squares and factors very happily:

$$\lim_{x \to -3} \frac{(x + 3)(x - 3)}{x + 3}$$

Now both the top and bottom of the fraction contain $(x + 3)$, so you can cancel those terms out to get the much simpler limit expression of:

$$\lim_{x \to -3}(x - 3)$$

Now you can use the substitution method to finish:

$$-3 - 3 = -6$$

So, $\lim\limits_{x \to -3} \frac{x^2 - 9}{x + 3} = -6$.

You've Got Problems

Problem 2: Evaluate these limits using the factoring method:

(a) $\lim\limits_{x \to 5} \dfrac{2x^2 - 7x - 15}{x - 5}$

(b) $\lim\limits_{x \to 1} \dfrac{x^3 - 1}{x - 1}$

Conjugate Method

If substitution and factoring don't work, you have one last bastion of hope when evaluating limits, but this final method is very limited in its scope and power. In fact, it is only useful for limits that contain radicals, as its power comes from the use of the conjugate.

The *conjugate* of a binomial expression (i.e., an expression with two terms) is the same expression with the opposite middle sign. For example, the conjugate of $\sqrt{x} - 5$ is $\sqrt{x} + 5$.

The true power of conjugate pairs is displayed when you multiply them together. The product of two conjugates containing radicals will, itself, contain no radical expressions! In other words, multiplying by a conjugate can eliminate square roots:

> **Talk the Talk**
>
> For our purposes, the **conjugate** of a binomial expression simply changes the sign between the two terms to its opposite. For example, $3 + \sqrt{x}$ and $3 - \sqrt{x}$ are conjugates.

$$\left(\sqrt{x} - 5\right)\left(\sqrt{x} + 5\right) = \sqrt{x^2} + 5\sqrt{x} - 5\sqrt{x} - 25 = x - 25$$

You should use the conjugate method whenever you have a limit problem containing radicals for which substitution does not work—always try substitution first. However, if substitution results in an illegal value $\left(\text{like } \frac{0}{0}\right)$, you'll know to employ the conjugate method, which we'll use to solve the next example.

Example 3: Evaluate $\lim\limits_{x \to 5} \dfrac{\sqrt{x+11} - 4}{x - 5}$.

Solution: If you try the substitution method, you get $\frac{0}{0}$, indicating that you'll need another method to find the limit since the function probably has a hole at $x = 5$. The function itself contains a radical and a number being subtracted from it—the fingerprint of a problem needing the conjugate method. To start, multiply both the numerator and denominator by the conjugate of the radical expression $\left(\sqrt{x+11} + 4\right)$:

$$\lim\limits_{x \to 5} \frac{\sqrt{x+11} - 4}{x - 5} \cdot \frac{\sqrt{x+11} + 4}{\sqrt{x+11} + 4}$$

Multiply the numerators and denominators as you would any pair of binomials—i.e., $(a + b)(c + d) = ac + ad + bc + bd$—and all of the radical expressions will disappear from the numerator. *Do not actually multiply the nonconjugate pair together.* You'll see why in a second:

$$\lim_{x \to 5} \frac{(x+11) + 4\sqrt{x+11} - 4\sqrt{x+11} - 16}{(x-5)(\sqrt{x+11}+4)}$$

$$= \lim_{x \to 5} \frac{x+11-16}{(x-5)(\sqrt{x+11}+4)}$$

$$= \lim_{x \to 5} \frac{x-5}{(x-5)(\sqrt{x+11}+4)}$$

Here's the neat trick: The numerator and denominator now contain the same term, so you can cancel that term and then finish the problem with the substitution method:

$$\lim_{x \to 5} \frac{1}{\sqrt{x+11}+4}$$

$$= \frac{1}{\sqrt{16}+4}$$

$$= \frac{1}{4+4} = \frac{1}{8}$$

You've Got Problems

Problem 3: Evaluate the following limits:

(a) $\lim\limits_{x \to -2} \dfrac{x+2}{\sqrt{x+6}-2}$

(b) $\lim\limits_{x \to 1} \dfrac{\sqrt{x+4}-3}{x+1}$

What If Nothing Works?

If none of the three techniques we have discussed works on the problem at hand, you're not out of hope. Don't forget we have an alternative (albeit tedious, mechanical, and unexciting—like most television sitcoms) method of finding limits. If all else fails, substitute a number insanely close to the number for which you are evaluating, as we did in Chapter 5.

Let me also play the part of the soothsayer for a moment. For the maximum effect, read the next sentences in a creepy, gypsy fortuneteller voice. "I see something in your future, yes, off in the distance. A promised method, a shortcut, a new way to evaluate limits that makes hard things easy. I'm getting a French name ... L'Hôpital's Rule, and an unlucky number ... 13. Chapter 13. Look for it in Chapter 13."

Limits and Infinity

There is a very deep relationship between limits and infinity. At first they thought they were "just friends," and then one would occasionally catch the other in a sidelong glance with eyes that spoke volumes. Without going into the long history, now they're insepara-ble, and without their storybook relationship, there'd be no vertical or horizontal asymp-totes.

Vertical Asymptotes

You already know that a limit does not exist if a function increases or decreases infinitely, like at a vertical asymptote. You may be wondering if it's possible to tell if a function is doing just that without having to draw the graph, and the answer is yes. Just like a substi-tution result of $\frac{0}{0}$ typically means a hole exists on the graph, a result of $\frac{5}{0}$ indicates a vertical asymptote. To be more specific, you don't have to get 5 in the numerator—any nonzero number divided by 0 indicates that the function is increasing or decreasing with-out bound, meaning no limit exists.

Example 4: At what value(s) of x does no limit exist for $f(x) = \dfrac{x^2 + 7x + 10}{x^2 - 25}$?

Solution: Begin by factoring the expression, because knowing what x values cause a 0 in the denominator is key:

$$f(x) = \frac{(x+5)(x+2)}{(x+5)(x-5)}$$

At $x = -5$, the function should have a hole, as substituting in that value results in $\frac{0}{0}$. You can use the factoring method to actually find that limit:

$$\lim_{x \to -5} f(x) = \frac{(-5+2)}{(-5-5)} = \frac{3}{10}$$

However, we're supposed to determine where the limit *doesn't* exist, so let's look at the other distressing x-value: $x = 5$. If you substitute that into f, you get $\frac{70}{0}$. This result, any number (other than 0) divided by 0, indicates the presence of a vertical asymptote at $x = 5$, so $\lim_{x \to 5} f(x)$ does not exist because f will either increase or decrease infinitely there.

If substitution results in $\frac{0}{0}$, that does not *guarantee* that a hole exists in the function. You can only be sure there's a hole there if a limit exists, as was the case with $x = -5$ in this example.

You've Got Problems

Problem 4: Describe the behavior of the function $g(x) = \dfrac{2x^3 - 3x^2 + x}{2x^3 + 5x^2 - 3x}$ at each x-value for which it is undefined.

Horizontal Asymptotes

Vertical asymptotes are caused by a function's values increasing or decreasing infinitely as that function gets closer and closer to a fixed x-value; so, if a function has a vertical asymptote at $x = c$, we can write $\lim_{x \to c} f(x) = \infty$ or $-\infty$. *Horizontal* asymptotes have a lot of the same components, but everything is reversed.

A horizontal asymptote is the height that a function tries to, but cannot, reach as the function's x-values get infinitely large or small. In Figure 6.2, f approaches a height of 5 as x gets infinitely large and a height of -1 as f grows infinitely negative.

Figure 6.2

The graph of f(x) *has different horizontal asymptotes as* x *gets infinitely positive and negative.*

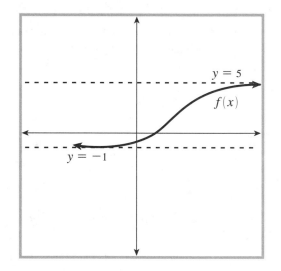

This is written as follows:

$$\lim_{x \to \infty} f(x) = 5 \text{ and } \lim_{x \to -\infty} f(x) = -1$$

These are called limits at infinity, since you are not approaching a fixed number, as you do with typical limits. However, just like the limits we've already discussed, the limit exists because the function clearly intends to reach the limiting height indicated by the horizontal asymptote, although it never actually reaches it.

Evaluating limits at infinity is a bit different from evaluating standard limits; substitution, factoring, and the conjugate methods won't work, so you need an alternative method. Although L'Hôpital's Method works quite nicely, you won't learn that until Chapter 13. In the meantime, you can evaluate these limits simply by comparing the highest exponents in their numerators and denominators.

Let's say we are trying to evaluate some function called $r(x)$, which is defined as a fraction whose numerator, $n(x)$, and denominator, $d(x)$, are simply polynomials. Compare the *degrees* (highest exponents) of $n(x)$ and $d(x)$:

- ◆ If the degree of the numerator is higher, then $\lim_{x \to \infty} r(x) = \infty$ or $-\infty$ (i.e., there is no limit because the function increases or decreases infinitely).

- ◆ If the degree of the denominator is higher, then $\lim_{x \to \infty} r(x) = 0$.

- ◆ If the degrees are the same, then $\lim_{x \to \infty} r(x)$ is equal to the *leading coefficient* of $n(x)$ divided by the leading coefficient of $d(x)$.

Critical Point

If $r(x)$ is a rational (fractional) function and has a horizontal asymptote, then it is guaranteed that $\lim_{x \to \infty} r(x) = \lim_{x \to -\infty} r(x)$. In other words, a rational function has the same horizontal asymptote as r goes to the left and the right sides of the graph.

These guidelines only apply to limits at infinity; make sure to remember that.

Example 5: Evaluate $\lim_{x \to \infty} \dfrac{5x^3 + 4x^2 - 7x + 4}{2 + x - 6x^2 + 8x^3}$.

Solution: This is a limit at infinity, so you should compare the degrees of the numerator and denominator. They are both 3, so the limit is equal to the leading coefficient of the numerator (5) divided by the leading coefficient of the denominator (8), so $\lim_{x \to \infty} \dfrac{5x^3 + 4x^2 - 7x + 4}{2 + x - 6x^2 + 8x^3} = \dfrac{5}{8}$.

Talk the Talk

The **degree** of a polynomial is the value of its largest exponent. The **leading coefficient** of a polynomial is the coefficient of the term with the largest exponent. For example, the expression $y = 3x^2 - 5x^6 + 7$ has degree 6 and leading coefficient –5.

You've Got Problems

Problem 5: Evaluate the following limits:

(a) $\lim_{x \to \infty} \dfrac{2x^2 + 6}{3x^2 - 4x + 1}$

(b) $\lim_{x \to -\infty} \dfrac{3x^2 + 4x + 3}{x^3 + 8x + 14}$

Special Limit Theorems

The four following special limits are not special because of the warm way they make you feel all giddy inside. By "special," I really mean they cannot be evaluated by the means we've discussed so far, but yet you'll see them frequently so you should probably memorize them, even though that stinks. Now that we're on the same page, so to speak, here they are with no further ado.

◆ $\lim\limits_{\alpha \to 0} \dfrac{\sin \alpha}{\alpha} = 1$

This is only true when you approach 0, so don't use this formula under any other circumstances. The α can be any quantity.

◆ $\lim\limits_{\alpha \to 0} \dfrac{\cos \alpha - 1}{\alpha} = 0$

Just like the first special limit, this is only true when approaching 0. Sometimes, you'll also see this formula written as $\dfrac{1 - \cos \alpha}{\alpha}$; the limit is still 0 either way.

◆ $\lim\limits_{x \to \infty} \dfrac{\text{any real number}}{x^{\text{any integer} \geq 1}} = 0$

If any real number is divided by x, and we let that x get infinitely large, the result is 0. Think about that—it makes good sense. What is 4 divided by 900 kajillion? Who knows, but it's definitely very, very small. So small, in fact, that it's basically 0.

◆ $\lim\limits_{x \to \infty} \left(1 + \dfrac{1}{x}\right)^x = e$

This basically says that 1 plus an extremely small number, when raised to an extremely high power, is exactly equal to Euler's number (2.71828 …). You will see this very infrequently, but it's important to recognize it when you do.

Example 6: Evaluate $\lim\limits_{x \to 0} \dfrac{\sin 3x}{x}$.

Solution: This is the first special limit formula, but notice that the value inside sine must match the denominator for that formula to work; therefore, we need a $3x$ in the denominator instead of just x. The trick is to multiply the top and bottom by 3 (since that's really the same thing as multiplying by 1, you're not changing the expression's value):

$$\lim\limits_{x \to 0} \dfrac{\sin 3x}{x} \cdot \dfrac{3}{3} = \lim\limits_{x \to 0} \dfrac{\sin 3x}{3x} \cdot 3$$

You can evaluate the limits of the factors separately and multiply the results together for the final answer:

$$\left(\lim_{x \to 0} \frac{\sin 3x}{3x}\right)\left(\lim_{x \to 0} 3\right) = (1)(3) = 3$$

Problem 6: Evaluate $\lim_{x \to \infty}\left(\frac{5}{x^3} + \left(1 + \frac{1}{x}\right)^x\right)$.

The Least You Need to Know

♦ Most limits can be evaluated via the substitution, factoring, or conjugate methods.

♦ If a function f has a vertical asymptote $x = c$, then $\lim_{x \to c} f(x) = \infty$ or $-\infty$.

♦ If a rational function f has a horizontal asymptote $y = L$, then $\lim_{x \to \infty} f(x) = \lim_{x \to -\infty} f(x) = L$.

♦ There are four common limits (two that approach 0 and two that approach ∞) that defy our techniques and must be memorized.

Chapter 7

Continuity

In This Chapter

- ◆ What it means to be continuous
- ◆ Classifying discontinuity
- ◆ When is the discontinuity removable?
- ◆ The Intermediate Value Theorem

Now that you understand and can evaluate limits, it's time to move forward with that knowledge. If you were to flip through any calculus textbook and read some of the most important calculus theorems, nearly every one contains a very significant condition: continuity. In fact, almost none of our most important calculus conclusions (including the Fundamental Theorem of Calculus, which sounds pretty darn important) work if the functions in question are not continuous.

Testing for continuity on a function is very similar to testing for the existence of limits on a function. Just as three stipulations had to be met in order for a limit to exist at a given point (left- and right-hand limits existing and being equal), three different stipulations must be met in order for a function to be continuous at a point. Just as there were three major cases in which limits did not exist, there are three major causes that force a function to be discontinuous.

Calculus is very handy like that—if you look hard enough, you can usually see clearly how one topic flows seamlessly into the next. Without limits, there'd be no continuity; without continuity, there'd be no derivatives; without derivatives, no integrals; and without integrals, no sleepless, panicked nights trying to cram for calculus tests.

What Does Continuity Look Like?

First of all, let's set our language straight. *Continuous* is an adjective that describes a function meeting very specific standards. Just like the Boy Scouts have to pass small tests to earn merit badges, there are three tests a function must pass at any given point in order to earn the "Continuous" merit badge.

Before we get into the nitty-gritty of the math definition, let's approach continuity from a visual perspective. It is easiest to determine whether or not a function is continuous by looking at its graph. *If the graph has no holes, breaks, or jumps, then we can rest assured that the function is continuous.* A continuous function is simply a nice, smooth function that can be drawn completely without lifting your pencil. With this intuitive definition in mind, in Figure 7.1, can you tell which of the following three functions is continuous?

Figure 7.1

One of these things is not like the others; one of these things just doesn't belong. Which is the continuous function?

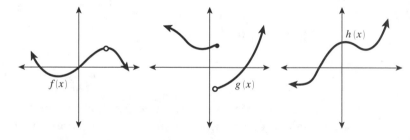

Critical Point

A continuous function is like a well-built roller coaster track—no gaps, holes, or breaks means safe riding for its passengers.

Of the above functions, $h(x)$ can be drawn with a single unbroken pen stroke. The other functions are much more unpredictable: f has an unexpected hole in it, and g suddenly breaks without warning. Only good old h provides a nice smooth ride from start to finish. Function h is like the good, solid, dependable boyfriend or girlfriend you wouldn't be embarrassed to bring home to Mom and Dad, and the fact that it guarantees no unexpected breakups means peace of mind for your emotional well-being.

The Mathematical Definition of Continuity

In his brilliant masterwork *Deep Thoughts*, humorist and philosopher Jack Handey provides the following plea: "Instead of having 'answers' on a math test, they should just call them 'impressions,' and if you got a different 'impression,' so what, can't we all be brothers?" Unfortunately, in math an "impression" is not enough. Math people are not touchy-feely like that. Very rarely, for instance, would you walk into the break room at an engineering company and find the engineers holding hands and singing "Give Peace a Chance."

Luckily, the mathematical definition of continuity makes a lot of sense if you keep one thing in mind: Whereas limits told us where a function intended to go, continuity guarantees that the function actually made it there. As the saying goes, "The road to hell is paved with good intentions." Continuity has the mathematical role of policeman, determining whether or not the function followed through with its intentions (meaning it is continuous) or not (making the function discontinuous). With that in mind, here is the official definition of continuity.

A function $f(x)$ is continuous at a point $x = c$ if the following three conditions are met:

- $\lim\limits_{x \to c} f(x)$ exists

- $f(c)$ is defined

- $\lim\limits_{x \to c} f(x) = f(c)$

In other words, the limit exists at $x = c$ (which means the function has an intended height); the function exists at $x = c$ (which means that there is no hole there); and the limit is equal to the function value (i.e., the function's value matches its intended value). (By the way, if a function is continuous, you can evaluate any limit on that function using the substitution method, since the function's value at any point will be equal to the limit there.)

Critical Point

A function is continuous at a point if the limit and function value there are equal. In other words, the limit exists if the intended height matches the actual function height.

Critical Point

Many functions are always guaranteed to be continuous at each point in their domain, including polynomial, radical, exponential, logarithmic, rational, and trigonometric functions. Most of the discontinuous functions you'll encounter will be due to undefined spots in rational functions and jumps due to piecewise-defined functions. We'll discuss more about specific causes of discontinuity in the next section.

Example 1: Show that the function:

$$f(x) = \begin{cases} \dfrac{\sqrt{x+19}-4}{x+3}, & x \neq -3 \\ \dfrac{1}{8}, & x = -3 \end{cases} \quad \text{is continuous at } x = -3.$$

Solution: To test for continuity, we must find the limit and the function value at $x = -3$ (and make sure they are equal). Now, that's one ugly function. How can we determine its intended height (limit) at $x = -3$? Clearly, the top rule in this piecewise-defined function governs the function's value for every single x except for $x = -3$. When you are finding a limit, you want to see what height is intended as you approach $x = -3$, not the value actually reached at $x = -3$, so we will find the limit of the larger, ugly, top rule for f. We'll need to use the conjugate method:

$$\lim_{x \to -3} \frac{\sqrt{x+19}-4}{x+3} \cdot \frac{\sqrt{x+19}+4}{\sqrt{x+19}+4}$$

$$= \lim_{x \to -3} \frac{(x+19)-16}{(x+3)(\sqrt{x+19}+4)}$$

$$= \lim_{x \to -3} \frac{x+3}{(x+3)(\sqrt{x+19}+4)}$$

$$= \lim_{x \to -3} \frac{1}{\sqrt{x+19}+4} = \frac{1}{\sqrt{16}+4} = \frac{1}{8}$$

The limit clearly exists when $x = -3$, and it is equal to $\frac{1}{8}$. The first condition of continuity is satisfied. Now, on to the second. According to the function's definition, we know that $f(-3) = \frac{1}{8}$, so the function does exist there. Finally, we can conclude that the function is continuous at $x = -3$, because the limit is equal to the function value there.

Types of Discontinuity

Not much happens in the life of a graph—it lives in a happy little domain, playing matchmaker to pairs of coordinates. However, there are three occurrences that can happen over the span of the function which change it fundamentally, making it discontinuous. Memorizing the three major causes of discontinuity is not so important; instead, recognize exactly what causes the function to fall short of continuity's requirements.

Jump Discontinuity

A *jump discontinuity* is typically caused by a piecewise-defined function whose pieces don't meet neatly, leaving gaping tears in the graph large enough for an elephant, or other tusked mammal, to walk through. Consider the function:

$$f(x) = \begin{cases} -x + 3, & x < 0 \\ x + 1, & x \geq 0 \end{cases}$$

This graph is made up of two linear pieces, and the rule governing the function changes when $x = 0$. Now look at the graph of f in Figure 7.2.

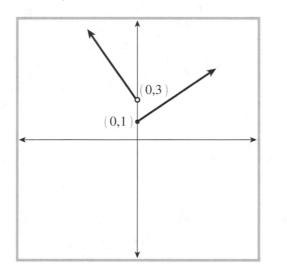

Figure 7.2

The graph of f(x) *exhibits an unhealthy dual personality since it is defined by a piecewise-defined function.*

When $x = 0$, the graph has a tragic and unsightly break. Whereas the left-hand piece is heading toward a height of $y = 3$, the right-hand piece starts from an entirely different height: $y = 1$. Does this sound familiar? It should: The left- and right-hand limits are unequal at $x = 0$, so $\lim_{x \to 0} f(x)$ does not exist. This breaks the first rule requirement of continuity, rendering f discontinuous.

Talk the Talk

A **jump discontinuity** occurs when no general limit exists at the given x value (because the right- and left-hand limits exist but are not equal).

In the next example, you are given a piecewise-defined function. Your goal is to shield it from the same fate as your pitiful function f above by choosing a value for the constant c, which ensures that the pieces of the graph will meet when the defining function rule changes. Godspeed!

Example 2: Find the real number, c, which makes $g(x)$ *everywhere continuous* if:

$$g(x) = \begin{cases} x - 2, & x \le 3 \\ x^2 + c, & x > 3 \end{cases}$$

Solution: The first rule in g will define the function for all numbers less than or equal to 3, and its reign ends once x reaches that boundary. At that point, g will have reached a height of $g(3) = 3 - 2 = 1$. Therefore, the next rule ($x^2 + c$) must start at *exactly* that height when $x = 3$, even though it is technically defined only when $x > 3$. That's the key: Both pieces must reach the exact same height when the graph of a piecewise-defined function changes rules. Thus, we know that:

Talk the Talk

A function is **everywhere continuous** if it is continuous for each x in its domain. Since g in Example 2 is made up of a linear piece and a quadratic piece (both of which are always continuous), the only place a discontinuity can occur is at $x = 3$.

$$x^2 + c = 1$$

when $x = 3$, so plug in that x value and solve for c:

$$3^2 + c = 1$$
$$9 + c = 1$$
$$c = -8$$

Thus, the second piece of g must be $x^2 - 8$ in order for g to be continuous. You can verify our solution with the graph of g (as shown in Figure 7.3)—no jump discontinuity anywhere to be found.

You've Got Problems

Problem 2: Find the value of a which makes the function $h(x)$ everywhere continuous if:

$$h(x) = \begin{cases} 2x^2 + x - 7, & x < -1 \\ ax + 6, & x \ge -1 \end{cases}$$

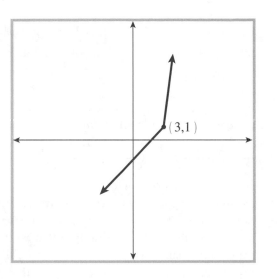

Figure 7.3
The graph of g(x) *is nice and continuous now—the pieces of the graph join together seamlessly.*

Point Discontinuity

A *point discontinuity* occurs when a function contains a hole. Think of it this way: The function is discontinuous only because of that rascally little point, hence the name.

Consider the function $p(x) = \dfrac{x^2 + 11x + 28}{x + 4}$. It is a rational function composed of two simple polynomials; we know that both of these functions will be everywhere continuous for all points on their domain. Hold on a second, though. The value $x = -4$ is definitely not in the domain of p (look at the denominator), so p will automatically be discontinuous there. The question is: What sort of discontinuity is present?

Classifying the discontinuity in this case is very easy—all you have to do is to test for a limit at that x value. To calculate the limit, use the factoring method:

$$\lim_{x \to -4} \frac{x^2 + 11x + 28}{x + 4}$$

$$= \lim_{x \to -4} \frac{(x + 7)(x + 4)}{x + 4}$$

$$= \lim_{x \to -4} (x + 7) = -4 + 7 = 3$$

Talk the Talk

A **point discontinuity** occurs when a general limit exists, but the function value is not defined there, breaking the second condition of continuity.

Critical Point

Any x-value for which a function is undefined will automatically be a point of discontinuity for the function. If a limit exists at the point of discontinuity, then it must be a point discontinuity.

In conclusion, since $x = -4$ represents a place where p is undefined, and $\lim\limits_{x \to -4} p(x) = 3$, we know that there is a hole in the function p at the point $(-4,3)$, a point discontinuity.

Infinite/Essential Discontinuity

An *infinite* (or *essential*) *discontinuity* occurs when a function neither has a limit (because the function increases or decreases without bound) nor is it defined at the given x-value. In other words, this type of discontinuity occurs primarily at a vertical asymptote.

Talk the Talk

Infinite discontinuity is caused by a vertical asymptote. Since the function increases or decreases without bound, there can be no limit, and since the function never actually touches the asymptote, the function is undefined there. Thus, the presence of a vertical asymptote ruins all the conditions necessary for continuity to occur. Vertical asymptotes are the home wreckers of the function world.

It's easy to determine which x-values cause a vertical asymptote, if you remember the shortcut from our previous work: A function increases or decreases infinitely at a given value of x if substituting that x into the expression results in a constant divided by 0. On the other hand, a result of $\frac{0}{0}$ *usually* means that point discontinuity is at work. However, since a result of $\frac{0}{0}$ doesn't *guarantee* that you've got point discontinuity, you'll need to double check to see if the limit exists there. We'll do this in Example 3, in case you're confused.

In summary: If no limit exists, you have jump discontinuity; if the limit exists but the function doesn't, you have point discontinuity; if the limit doesn't exist because it is ∞ or $-\infty$, you have infinite discontinuity.

Now that you've got the field guide to discontinuity, let's look at a typical problem you'll be given. In it, you'll be asked either to identify when a function is continuous or instead to highlight areas of discontinuity and to classify the type of discontinuity present.

Example 3: Give all x-values for which the function:

$$f(x) = \frac{9x^2 - 3x - 2}{3x^2 + 13x + 4}$$

is discontinuous, and classify each instance of discontinuity.

Solution: This is a rational function comprised of quadratic functions; each piece is guaranteed to be continuous on its entire domain, so the only points we have to inspect are where f is undefined. Because f is rational, it is undefined when its denominator equals 0, and the easiest way to find those locations is by factoring f:

$$f(x) = \frac{(3x + 1)(3x - 2)}{(x + 4)(3x + 1)}$$

Set the denominator equal to 0 to see that $x = -4$ and $x = -\frac{1}{3}$ will be points of discontinuity. Now, we need to explain what kinds of discontinuity they are. Plug each into f. Substituting $x = -4$ results in $\frac{154}{0}$, indicating the presence of a vertical asymptote and an infinite discontinuity. However, substituting $x = -\frac{1}{3}$ into f gives you $\frac{0}{0}$, which means there is probably a hole in the function there. That's not good enough supporting work, however. You need to *prove* that there's a hole there in order to conclude that $x = -\frac{1}{3}$ represents a point discontinuity. All the proof you'll need is to verify the presence of a limit at $x = -\frac{1}{3}$. To do so, use the factoring method:

$$\lim_{x \to -1/3} \frac{(3x+1)(3x-2)}{(x+4)(3x+1)}$$

$$= \lim_{x \to -1/3} \frac{3x-2}{x+4} = \frac{3\left(-\frac{1}{3}\right)-2}{-1/3+4} = \frac{-1-2}{11/3} = -\frac{9}{11}$$

Because the limit exists, there is a point discontinuity when $x = -\frac{1}{3}$.

You've Got Problems

Problem 3: Give all x-values for which the function …

$$g(x) = \frac{2x^2 + 5x - 25}{x^2 - 25}$$

… is continuous, and classify each instance of discontinuity.

Removable vs. Nonremovable Discontinuity

Occasionally you'll see a function described as having removable or nonremovable discontinuity. These terms are more specific than simply stating that a function is discontinuous, but less specific than the kinds of discontinuity we explored in the last section. In other words, these terms are vague, whereas point, jump, and infinite discontinuity are not.

However, since these terms appear often, it's good to know what they mean. A *removable discontinuity* is one that could be eliminated by simply redefining a finite number of points in the function. In other words, if you can "fix" the function by filling in holes, then that function is removably discontinuous. Let's go back to the function we described in Example 3 for a second:

$$f(x) = \frac{(3x + 1)(3x - 2)}{(x + 4)(3x + 1)}$$

This function has a point discontinuity at $x = -\frac{1}{3}$, since $\lim_{x \to -1/3} f(x) = -\frac{9}{11}$. If I choose, I can change the definition of f to make it continuous. If I make $f\left(-\frac{1}{3}\right) = -\frac{9}{11}$, then the limit equals the function value there, and f becomes continuous. Mathematically, the new function f looks like this:

$$f(x) = \begin{cases} \dfrac{(3x + 1)(3x - 2)}{(x + 4)(3x + 1)}, & x \neq -\dfrac{1}{3} \\ -\dfrac{9}{11}, & x = -\dfrac{1}{3} \end{cases}$$

You don't actually have to change the function for it to be removably discontinuous (in fact, if you did change the function, it wouldn't be discontinuous at all). However, if it is *possible* to change a few points in order to fill in the function's holes, the function's discontinuities are removable.

Nonremovable discontinuity occurs when a function has no general limit at the given x-value, as is the case with infinite and jump discontinuities. There is no way to redefine a finite number of points to "repair" this type of discontinuous function. This function is fundamentally discontinuous, and no amount of rehabilitation or mood-altering medication can make this function safe for a cultured society. I gasp at the thought, but not even a charity rock concert can help (sorry, U2). Back to the function f from Example 3 for illustration. Since a vertical asymptote occurs at $x = -4$ and no general limit exists there, $x = -4$ represents an instance of nonremovable discontinuity.

Talk the Talk _____

A function is **removably discontinuous** if it only contains point discontinuity, since you can redefine the function to fill in the holes (and thus *remove* the discontinuity) if you chose to. For a point of discontinuity to be removable, a limit must exist there. Therefore, point and removable discontinuity are essentially synonymous. A function is **nonremovably discontinuous** at a given x-value if no general limit exists there, making it impossible to remove the discontinuity by redefining a fixed number of points. Jump and infinite discontinuities are both examples of nonremovable discontinuity.

The Intermediate Value Theorem

Break out the party favors—we've arrived at our first official calculus theorem.

The Intermediate Value Theorem: If a function $f(x)$ is continuous on the closed interval $[a,b]$, then for every real number d between $f(a)$ and $f(b)$, there exists a c between a and b such that $f(c) = d$.

Now, let me explain what the heck that means using a simpler example. Like all red-blooded Americans, I enjoy a little too much holiday dining during the winter months of November and December. If we were to exaggerate my weight gain (a little), it might have the following humorously titled "Date vs. Weight" graph, which in Figure 7.4 I'll call $w(x)$.

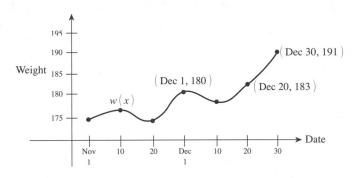

Figure 7.4

Kelley's Date vs. Weight graph.

From the graph, we can see that I weighed 180 pounds on December 1 and porked up to 191 by the time December 30 "rolled around," poor pun definitely intended. Comparing this to the Intermediate Value Theorem, a = Dec 1, b = Dec 30 (weird values, but go with me on this), $f(a)$ = w(Dec 1) = 180, and $f(b)$ = w(Dec 30) = 191. According to the theorem, I can choose any value between 180 and 191 (for example, 183), and I am guaranteed that at some time between December 1 and December 30, I actually weighed that much.

The Intermediate Value Theorem does not claim to tell you *where* your function reaches that value or *how many times* it does. The theorem simply claims (in a calm, soothing voice) that every height a function reaches on a specific x-interval boundary will be the output at least once by some x within that interval. As it only guarantees the existence of something, it is called an existence theorem.

You've Got Problems

Problem 4: Use the Intermediate Value Theorem to explain why the function $g(x) = x^2 + 3x - 6$ *must* have a root (x-intercept) on the closed interval $[1,2]$.

The Least You Need to Know

- A continuous function has no holes, jumps, or breaks in its graph.
- If a function reaches its intended height at a particular *x*-value, the function is continuous there.
- If a function is undefined but possesses a limit, you have point discontinuity, which is removable.
- Infinite discontinuity is caused by a vertical asymptote, whereas a jump discontinuity is caused by a break in the function's graph; both are nonremovable.
- The Intermediate Value Theorem uses rather complex language to guarantee the "wholeness" or "completeness" of a graph.

Chapter 8

The Difference Quotient

In This Chapter

◆ Creating a tangent from scratch
◆ How limits can calculate slope
◆ "Secant you shall find" the tangent line
◆ Both versions of the difference quotient

Although limits are important to the development of calculus and are the only topic we have even discussed so far, they are about to take a backseat to the two major topics comprising what most people call "calculus": derivatives and integrals. It would be rude (and actually mathematically inaccurate) to simply start talking about derivatives without describing their relationship to limits.

Brace yourself. This chapter describes the solution to one of the most puzzling mathematical dilemmas of all time: how to calculate the slope of a tangent line to a nonlinear function. We're going to use limits to concoct a general formula that will allow you to find the tangent slope to a function at any given point. The process is a little tedious and is a bit algebra-intensive. You may ask yourself, "Am I always going to go through so much pain to find a derivative?" The answer is no. In Chapter 9, you'll learn lots of shortcuts to finding derivatives.

For now, however, prepare to be dazzled. You're about to create a tangent line to a function and calculate its slope through a little "mathemagics."

Furthermore, the formula we're going to create is pretty complex, but you're going to understand exactly where it comes from and how it works. Even though this is a very theoretical process, it's one of the neatest theorems in calculus, and I'm not just saying that because I'm a math geek, either. You may not harbor the same feelings of love for it that I do, but you'll probably at least find it moderately attractive.

When a Secant Becomes a Tangent

Before we go about calculating the slope of a tangent line, you should probably know what a tangent line is. A *tangent line* is a line that just barely skims across the edge of a curve, hitting it at the point you intend.

In Figure 8.1, you'll see the graph of $y = \sin x$ with two of its tangent lines drawn, one at $x = \frac{\pi}{2}$ and one at $x = \frac{7\pi}{4}$. Notice that the tangent lines barely skim across the edge of the graph and hit only at one point, called the *point of tangency*. If you extend it, the tangent line may hit the graph again somewhere else along the graph, but that doesn't matter. What matters is that it only hits once relatively close to the point of tangency.

Figure 8.1

The point of tangency.

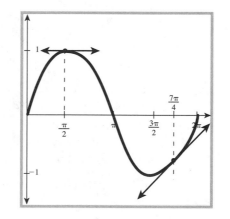

A *secant line*, on the other hand, is a line that crudely hacks right through a curve, usually hitting it in at least two places. In Figure 8.2, I have drawn both a secant and a tangent line to a function $f(x)$ when $x = 3$. Notice that the dotted secant line doesn't have the finesse of the tangent line, which strikes only at $x = 3$.

Through a little trickery, we are going to make a secant line into a tangent. This is the backbone of our procedure for calculating the slope of a tangent line. So now that you know what the words mean, let's get started.

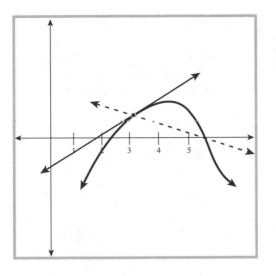

Figure 8.2

*A secant and a tangent line
to a function* f(x) *when*
x = 3.

Honey, I Shrunk the Δx

Take a look at the function graph in Figure 8.3
called $f(x)$. I have marked the location $x = c$ on
the graph. My final, overall goal will be to calculate the slope of the tangent line to f at $x = c$. You
may not understand this to be a very important
goal, but trust me, it is world-shatteringly important.

Talk the Talk

A **tangent line** skims
across the curve, hitting it once
in the indicated location; however, a secant line does not skim
at all. It cuts right through a function, usually intersecting it in multiple spots.

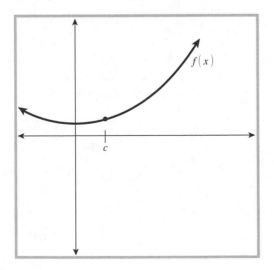

Figure 8.3

The graph of some function
f(x) *with location* x = c.

Now, let's add a few things to the graph to create Figure 8.4. First of all, I know the coordinates of the indicated point. In order to get to that point from the origin, I have to go c units to the right and $f(c)$ units up (so that I hit the function), which translates to the coordinate pair of $(c, f(c))$. Now let's add another point to the graph to the right of the point at $x = c$. How far to the right, you ask? Let's be generic and call it "Δx" more to the right. ("Δx" is math language for "the change in x," and since we're changing the x value of c by going Δx more to the right, it's a fitting name.)

Figure 8.4

Now appearing on the graph of f(x), *a new* x *value, which is a distance of* Δx *away from* c.

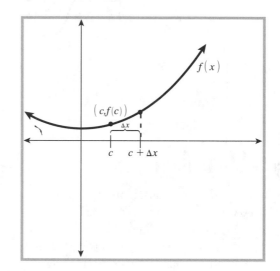

Once again, all we're doing is making a new point that is a horizontal distance of Δx away from the first point. Can you figure out the coordinates of the new point? In the same fashion that we got the first coordinate pair to be $(c, f(c))$, this point has coordinates $(c + \Delta x, f(c + \Delta x))$. Now connect these two points together, and what have you got? A secant line through f, as pictured in Figure 8.5.

True, our final goal is to find the slope of the tangent line to f at $x = c$, but for now, we'll amuse ourselves by finding the slope of the secant line we've drawn at $x = c$. We know how to calculate the slope of a line if given two points—use the procedure from Problem 3 in Chapter 2:

$$\text{slope} = \frac{y_2 - y_1}{x_2 - x_1}$$
$$= \frac{f(c + \Delta x) - f(c)}{c + \Delta x - c}$$
$$= \frac{f(c + \Delta x) - f(c)}{\Delta x}$$

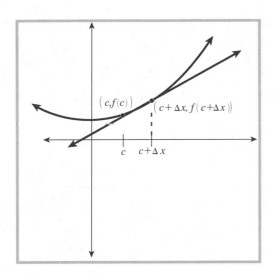

Figure 8.5

*We've added the coordinates
to the new point.*

So, that wasn't so bad. We found the slope of
the secant line, and that slope is very close to
the slope of the tangent line we want to find—
both have nearly the same incline. However, we
don't want an approximation of the slope of the
tangent line, we want it *exactly*. Here's the key: I
am going to redraw the second point on the graph
of f (remember the one that was Δx away from the
first point), and this time I am going to make Δx
smaller. Figure 8.6 shows this new, improved
secant line. Why is it improved? It has a slope
closer to the tangent line we're searching for.

Critical Point

Here comes the connection
to limits: The smaller I make
Δx, the closer the slope of
the secant line approxi-
mates the slope of the tangent line. I am not allowed to
make $\Delta x = 0$, because that
would mean I was dividing by 0
in the slope equation we created
a few moments ago.

Figure 8.6

*When Δx is smaller, the new
point is closer to x = c.*

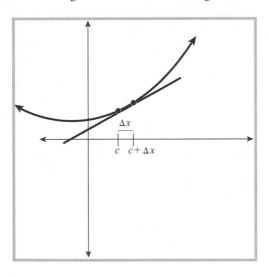

This new secant line is slightly shallower than the previous one, and it is an even better impersonator of the actual tangent line at $x = c$. The funny thing is, if I were to calculate its slope, it would look exactly the same as the slope I came up with before: $\dfrac{f(c + \Delta x) - f(c)}{\Delta x}$.

Here's my moment of brilliance: If I make Δx infinitely small, so small that it is basically (but not quite) 0, then the two points on the graph would be so close together that I would, in effect, actually have the tangent line. Therefore, by calculating the secant line slope, I'd actually be calculating the tangent line slope as well.

How do I make Δx get that small, though? It's easy, actually, since we know limits. We're just going to find the limit of the secant slope function *as Δx approaches 0*. This limit is called the *difference quotient*, and is the very definition of the *derivative*:

$$\lim_{\Delta x \to 0} \frac{f(c + \Delta x) - f(c)}{\Delta x}$$

Critical Point

The formula is called the difference quotient because (1) it represents a quotient since it is a fraction, and (2) the numerator and denominator represent the difference in the *y*'s and the *x*'s, respectively, between the two points on our secant line.

This is the most important calculus result we have discussed thus far. We now have an admittedly ugly but very functional formula allowing us to calculate the slope of the tangent line for a function. What's amazing is that we really forced it, didn't we? We actually created a tangent line out of thin air by forcing a secant line to undergo radical and mind-altering changes. But if you're like me, you're thinking, "Enough with the theory already—I'll probably never have to create the definition of a derivative. Instead, I'd rather know how to use the difference quotient to find a derivative." Your wish is my command.

Talk the Talk

The **derivative** of a function $f(x)$ at $x = c$ is the slope of the tangent line to f at $x = c$. You can find the value of the derivative using the **difference quotient,** which is this formula:

$$\lim_{\Delta x \to 0} \frac{f(x + \Delta x) - f(x)}{\Delta x}$$

As you can see, I usually write the formula with *x*'s instead of *c*'s, but that doesn't change the way it works.

Applying the Difference Quotient

In order to find the derivative $f'(x)$ of the function $f(x)$, simply follow the difference quotient formula. To get the numerator, you'll plug $(x + \Delta x)$ into f and then subtract the original function $f(x)$. Then, divide that quantity by Δx, and calculate the limit of the entire fraction as Δx approaches 0.

Example 1: Use the difference quotient to find the derivative of $f(x) = x^2 - 3x + 4$, and then evaluate $f'(2)$.

Solution: The difference quotient has one ugly piece in the numerator: $f(x + \Delta x)$, so let's figure out exactly what that is ahead of time and then plug it into the formula. Remember, when you evaluate $f(x + \Delta x)$, you have to plug in $x + \Delta x$ into *both* x terms in f. In other words, plug it into x^2 *and* $-3x$:

Kelley's Cautions

The most common errors people make when applying the difference quotient are forgetting to subtract $f(x)$ in the numerator and omitting the denominator completely. Sometimes evaluating $f(x + \Delta x)$ gets so tedious they forget the rest of the formula.

$$f(x + \Delta x) = (x + \Delta x)^2 - 3(x + \Delta x) + 4$$
$$= x^2 + 2x\Delta x + (\Delta x)^2 - 3x - 3\Delta x + 4$$

That entire disgusting quantity must be substituted into the difference quotient for $f(x + \Delta x)$ now, and we'll try to simplify as much as possible:

$$\lim_{\Delta x \to 0} \frac{f(x + \Delta x) - f(x)}{\Delta x}$$

$$= \lim_{\Delta x \to 0} \frac{\left(x^2 + 2x\Delta x + (\Delta x)^2 - 3x - 3\Delta x + 4\right) - \left(x^2 - 3x + 4\right)}{\Delta x}$$

$$= \lim_{\Delta x \to 0} \frac{x^2 - x^2 + 2x\Delta x + (\Delta x)^2 - 3x + 3x - 3\Delta x + 4 - 4}{\Delta x}$$

$$= \lim_{\Delta x \to 0} \frac{2x\Delta x + (\Delta x)^2 - 3\Delta x}{\Delta x}$$

You've got to admit—that looks a lot better than it did a second ago. You were starting to panic, weren't you? All of these difference quotient problems are going to simplify significantly like this. Now, how do we evaluate the limit? Substitution is a no-go, because it

Critical Point _____

There are many notations that indicate a derivative. The most common are: $f'(x)$, y', and $\frac{dy}{dx}$. The last two of these are typically used when the original function is written in "$y =$" form, rather than "$f(x)=$" form. The second derivative (the derivative of the first derivative) is denoted $f''(x)$, y'', and $\frac{d^2 y}{dx^2}$.

results in $\frac{0}{0}$, so we should move on to the next available technique: factoring. That works like a charm:

$$\lim_{\Delta x \to 0} \frac{\Delta x(2x + \Delta x - 3)}{\Delta x}$$

$$= \lim_{\Delta x \to 0} (2x + \Delta x - 3) = 2x + 0 - 3$$

$$f'(x) = 2x - 3$$

Now for the second part of the problem: calculating $f'(2)$, the slope of the tangent line when $x = 2$. It's as easy as plugging $x = 2$ into the newfound derivative formula:

$$f'(x) = 2x - 3$$
$$f'(2) = 2(2) - 3 = 1$$

You've Got Problems

Problem 1: Find the derivative of $g(x) = 5x^2 + 7x - 6$ and use it to calculate $g'(-1)$.

Once you find the general derivative using the difference quotient ($f'(x) = 2x - 3$), you can then calculate any specific derivative you desire (like $f'(2)$). However, finding that general derivative is not a whole lot of fun. In fact, it's just about as fun as that time you got nothing but socks and underpants for your birthday. Calculus does offer you an alternative form of the difference quotient if you feel hatred toward this method welling up inside of you.

The Alternate Difference Quotient

I have good news and bad news for you. First, the good news: This alternative derivative technique involves much less algebra and absolutely no Δx's at all. But, the bad news is that it cannot find the general derivative—you can only calculate specific values of the derivative. In other words, you'll be able to use this method to find values such as $f'(3)$, but you won't be able to find the actual derivative $f'(x)$. This definitely limits the usefulness of the alternative, but it is, without question, much faster than the first method, once you get used to it.

The alternate difference quotient: The derivative of f at the specific x-value $x = c$ can be found using the formula:

$$f'(c) = \lim_{x \to c} \frac{f(x) - f(c)}{x - c}$$

Notice the major differences between this and the previous difference quotient. For one thing, in this limit you approach the number x at which you are finding the derivative; in the other method, Δx always approached 0. In the numerator of this formula, you will calculate $f(c)$, which will be a real number; in the previous formula, both pieces of the numerator, $f(x + \Delta x)$ and $f(x)$, were functions of x. Clearly, the two formulas have different denominators as well. Since both are limits, though, evaluating them is quite similar once you've plugged in the initial values.

For grins, let's redo the second part of Example 1, since we already know the correct answer. You ever notice that math teachers just *love* doing this—reworking the same problem twice using different methods and arriving at the same answer as if by magic? I remember doing this in class, turning around at the conclusion of the second problem, and saying, "You see, they're equal!" Needless to say, I was the only one impressed. However, I still do this, hoping against hope that one day a student will faint from pure shock and delight when the answers work out the same.

Example 2: Evaluate $f'(2)$ if $f(x) = x^2 - 3x + 4$.

Solution: The alternative formula requires us to know $f(c)$, in this case $f(2)$, so I usually calculate that first:

$$f(2) = 2^2 - 3 \cdot 2 + 4 = 4 - 6 + 4 = 2$$

Now, plug that into the alternate difference quotient, and you'll be pleasantly surprised how much simpler it looks than Example 1:

$$f'(2) = \lim_{x \to 2} \frac{f(x) - f(2)}{x - 2}$$

$$= \lim_{x \to 2} \frac{\left(x^2 - 3x + 4\right) - (2)}{x - 2}$$

$$= \lim_{x \to 2} \frac{x^2 - 3x + 2}{x - 2}$$

To finish, evaluate the limit using the factoring method:

$$\lim_{x \to 2} \frac{(x-2)(x-1)}{(x-2)}$$

$$\lim_{x \to 2} (x-1) = 1$$

Like magic (although I am sure you're unimpressed), we get the same answer as before. Ta-da!

You've Got Problems

Problem 2: Calculate the derivative of $h(x) = \sqrt{x+1}$ when $x = 8$ using the alternate difference quotient.

The Least You Need to Know

- The slope of the tangent line to a curve at a certain point is called the derivative at that point.
- There are two forms of the different quotient; both give the value of a function's derivative.
- The original form of the difference quotient can provide the general derivative formula for a function, whereas the second can only give the derivative's value at a specified x-value.

Part 3

The Derivative

At the end of Part 2, you learned the basics of the difference quotient and that it calculates something called the derivative. In the study of calculus, the derivative is *huge*. Just about everything you do from here on out is going to use derivatives to some degree. Therefore, it's important to know exactly what they are, when they do and don't exist, and how to find derivatives of functions.

Once you've got the basic skills down, you can begin to explore the huge forest of applications that comes along with the derivative package. Since derivatives are actually rates of change, they classify and describe functions in ways you'll hardly believe. Have you ever wondered, "What's the maximum area I could enclose with a rectangular fence if one side of the rectangle is three times more than twice the other side?" If you have, well, you scare me because no one has thoughts like that. However, the good news is that you'll be able to find your answer once and for all.

Laying Down the Law for Derivatives

In This Chapter

◆ When can you find a derivative?
◆ Calculating rates of change
◆ Simple derivative techniques
◆ Derivatives of trigonometric functions
◆ Multiple derivatives

One of my most memorable college professors was a kindly Korean man named Dr. Oh. One of the reasons his class sticks out in my mind is the way he was able to illustrate things with bizarre but poignant imagery. The day we first discussed the Fundamental Theorem of Calculus, he described it in his usual, understated way. "Today's topic is like the day the world was created. Yesterday, not interesting. Today, interesting!"

One of the classes I took from Dr. Oh was Differential Equations. Dr. Oh constantly (but jokingly) harassed the young lady who sat next to me, because she would always do things the long way. No matter what shortcuts we learned, she wouldn't use them. I never understood why, and she simply explained to me, "This is the way I do things …. I can't change it now!" I remember Dr. Oh repeatedly asking her, "If you want potatoes, do you buy a farm, till the field, plant the seed, nurture the plants, and then harvest the potatoes? If I were you, I would just go to the grocery store."

In the land of derivatives, the difference quotient is the equivalent of growing your own potatoes. Sure, the process works, but I gave you very specific examples so that it would work for you without any trouble or heartache. I was shielding you against the harsh weather of complicated derivatives to come. However, I have to let you grow up some-time and stare in the face of an ugly, complicated derivative. The good news, though, is that you can buy all your solutions from the grocery store.

When Does a Derivative Exist?

Before you run around finding derivatives willy-nilly, you should know that there are three specific instances in which the derivative to a function fails to exist. Even if you get a numerical answer when calculating a derivative, it's possible that the answer is invalid, because there actually is no derivative! Be extra cautious if the graph of your function contains any of the following things:

> **Critical Point**
>
> You'll hear the statement "Differentiability implies continuity" in your calculus class. That means exactly this: If a function has a derivative at a specific x-value, then the function *must* also be continuous at that x value. That statement is the logical equiva-lent of saying, "If a function is *not* continuous at a certain point, then that function is not differen-tiable there either."

Discontinuity

A derivative cannot exist at a point of discontinuity. It doesn't matter if the discontinuity is removable or not. If a function is discontinuous at a specific x-value, there cannot be a derivative there. For example, if you are given the function …

$$f(x) = \frac{(x-1)(x+2)}{(x+2)(x-6)}$$

… you know, without a bit of work, that f has no deriv-ative at $x = -2$ and $x = 6$. In other words, f is not *differentiable* at those values of x.

Sharp Point in the Graph

If a graph contains a sharp point (also known as a cusp), then the function has no deriva-tive at that point. Not many functions have cusps; in fact, they are pretty rare. You're most likely to see them in functions containing absolute values and in piecewise-defined functions whose pieces meet, but not smoothly. In Figure 9.1, you'll find the graphs of the functions …

$$f(x) = |x - 1| - 2 \text{ and } g(x) = \begin{cases} x^2, x \le 1 \\ x, x > 1 \end{cases}$$

... which both contain nondifferentiable cusps at $x = 1$.

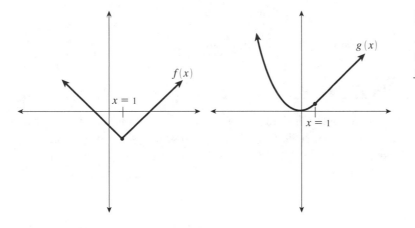

Figure 9.1

Both graphs are pointy at x = 1 and, therefore, nondifferentiable there.

Kelley's Cautions

Many modern calculators can evaluate derivatives, but a lot of them will give you incorrect derivative values if you are trying to find the derivative at a cusp. For example, if you differentiate $h(x) = |x|$, many calculators will tell you that $h'(0) = 0$, but we know that $h(x)$ has a cusp at $x = 0$, so there is really no derivative at all! Therefore, take any answer your calculator gives you with a grain of salt; it expects you to know if you are asking it to perform an impossible task.

Vertical Tangent Line

Remember that the derivative is defined as the slope of the tangent line. What if the tangent line is vertical? Keep in mind that vertical lines are described as having no slope at all. So, you can't find a derivative if the tangent line happens to be vertical. It's pretty tough to spot when this happens on a graph, but luckily, the mathematics of derivatives is quick to expose it when it happens, as shown in the next example.

Talk the Talk

A function is **differentiable** at a given value of x if you can take the derivative of the function at that x value. In other words, $f(x)$ is differentiable at $x = c$ if $f'(c)$ exists. A function whose derivative does not exist at a specific x value is said to be **nondifferentiable** there.

Example 1: Show that no derivative exists for the function $f(x) = x^{1/3}$ when $x = 0$.

Solution: You don't know how to find the derivative of $f(x) = x^{1/3}$ yet (but you will soon), so I'll tell you that it's $f'(x) = \frac{1}{3x^{2/3}}$. If you try to evaluate $f(0)$, you get:

$$f'(0) = \frac{1}{3(0)^{2/3}} = \frac{1}{0}$$

The slope of the tangent line is a nonexistent number, since you can't divide by 0.

Basic Derivative Techniques

Learning how to find derivatives using the difference quotient can be long, tedious work, but now that you've learned it, you've "paid your dues," so to speak. Now, you'll learn some really handy techniques. First of all, the Power Rule is going to help you find basic derivatives very easily, and the three rules following it will help you find derivatives of more complicated functions.

The Power Rule

Even though the Power Rule can only find very basic derivatives, you'll definitely use it more than any other of the rules we'll learn. In fact, it often pops up in the final steps of those rules, but let's not get ahead of ourselves. Any term in the form ax^n can be differentiated using the Power Rule.

The Power Rule: The derivative of the term ax^n (with respect to x), where a and n are real numbers, is $(a \cdot n)x^{n-1}$.

Critical Point

Don't worry about the phrase "with respect to x" in the Power Rule definition. Since x is the only variable in the expression, we really don't need to say that. In Chapter 10 we'll differentiate *implicitly*, and then you'll have to know what that actually means. For now, just understand that the phrase "with respect to" will refer to the variable in the problem, but won't affect any of your derivative techniques.

In simpler terms, this is what the Power Rule says (and with a name like the *Power Rule*, I'd listen if I were you): If you have a variable being raised to an exponent (whether it has a coefficient or not), you can find its derivative by the Power Rule, first taking the exponent, and multiplying it by the variable's coefficient. The result will be the derivative's coefficient. Second, subtract 1 from the exponent—the result is the power of the derivative. Some examples will shed some light on the matter.

Example 2: Use the Power Rule to find the derivative of $f(h) = \frac{4}{3}h^3 + 6h - 5$.

Solution: Even though there are a number of terms here, we can find the derivative of each using the Power Rule. Before we start, I'll tell you that the derivative of the constant term (–5) is 0. (See below for an explanation.) For the other terms, multiply the coefficient by the exponent and then subtract 1 from the exponent for each term separately:

$$f'(h) = \left(\frac{4}{3} \cdot 3\right)h^{3-1} + (6 \cdot 1)h^{1-1} - 0$$
$$= 4h^2 + 6$$

Remember, the exponent of $6h$ is understood to be 1 since it's not written.

The derivative of any constant is 0. Here is a quick justification if you are interested. Consider the constant function $g(x) = 7$. If we wanted to, we could write this function with a variable term: $g(x) = 7x^0$. I am not changing the value of the function since a variable to the 0 power has a value of 1, and multiplying $7 \cdot 1$ does not change the value of 7. So now that we've rewritten the function, use the Power Rule:

$$g'(x) = (7 \cdot 0)x^{0-1}$$
$$= 0x^{-1} = 0$$

You've Got Problems

Problem 1: Find derivatives using the Power Rule:

(a) $y = \frac{2}{3}x^3 + 3x^2 - 6x + 1$

(b) $f(x) = \sqrt[3]{x} + 2\sqrt[5]{x}$

The Product Rule

If a function contains two variable pieces being multiplied, you cannot simply find the derivative of each and multiply the results. For example, the derivative of $x^2 \cdot (x^3 - 3)$ is *not* $(2x)(3x^2)$. Instead, you have to use a very simple formula, which (by the way) you should memorize.

The Product Rule: If a function $h(x)$ is the product of two differentiable functions $f(x)$ and $g(x)$ …

$$h(x) = f(x) \cdot g(x)$$

… then the derivative of h is given by:

Kelley's Cautions

Overlooking the Product Rule is a very common mistake in calculus. Remember: If two variable expressions are being multiplied together, you *have* to use the Product Rule. If, however, you want to find the derivative of $5 \cdot 7x^2$, you don't need the Product Rule (since 5 is not a variable expression). Instead, you can rewrite it as $35x^2$ and get the correct derivative of $70x$.

$$b'(x) = f(x) \cdot g'(x) + f'(x) \cdot g(x)$$

Here's what that means. If a function is created by multiplying two other functions together, then the derivative of the overall function is the first times the derivative of the second plus the second times the derivative of the first. It's actually a very simple procedure, so let's get right into the examples.

Example 3: Differentiate the function $f(x) = (x^2 + 6)(2x - 5)$ using (1) the Product Rule, and (2) the Power Rule, and show that the results are equal. *Hint:* To use the Power Rule, you'll first have to multiply the terms together.

Solution: (1) The derivative, according to the Product Rule, is …

$$f'(x) = \left(x^2 + 6\right)(2) + (2x)(2x - 5)$$
$$= 2x^2 + 12 + 4x^2 - 10x$$
$$= 6x^2 - 10x + 12$$

(2) As the hint indicates, we need to multiply those binomials together before we can apply the Power Rule. Doing so results in $f(x) = 2x^3 - 5x^2 + 12x - 30$. Now, apply the Power Rule to get $f'(x) = 6x^2 - 10x + 12$, which matches the previous answer.

> **You've Got Problems**
>
> Problem 2: Find the derivative of $g(x) = (2x - 1)(x + 4)$ using (1) the Power Rule, and (2) the Product Rule, and show that the results are the same.

The Quotient Rule

Just as the Product Rule prevents you from simply taking individual derivatives when you're multiplying, the Quotient Rule prevents the same for division. Every year on my first derivatives exam, one of the problems is always to find the derivative of something like $\dfrac{x^2 + 7x}{3x^3 + 2x + 4}$, and half of my students always respond with an answer of $\dfrac{2x + 7}{9x^2 + 2}$, no matter how many times I warned them to use the Quotient Rule. You *must* use the Quotient Rule any time two variable expressions are being divided.

The Quotient Rule: If $h(x) = \frac{f(x)}{g(x)}$, where $f(x)$ and $g(x)$ are two differentiable functions,

then $h'(x) = \dfrac{g(x) \cdot f'(x) - f(x) \cdot g'(x)}{\left(g(x)\right)^2}$.

In other words, to find the derivative of a fraction, take the bottom times the derivative of the top and subtract the top times the derivative of the bottom; divide all of that by the bottom squared. Of course, by top and bottom, I mean numerator and denominator, respectively.

Example 4: Find the derivative of $y = \dfrac{3x+7}{x^2-1}$ using the Quotient Rule.

Solution: In the formula for Quotient Rule, $f(x) = 3x + 7$ and $g(x) = x^2 - 1$. Therefore, $f'(x) = 3$ and $g'(x) = 2x$. Plug all of these values into the appropriate spots in the Quotient Rule to get:

Kelley's Cautions

It is very important to get the subtraction order in the numerator of the Quotient Rule correct. Whereas in the Product Rule, either of the two functions could be f or g, in the Quotient Rule, g must be the denominator of your original function.

$$y' = \frac{\left(x^2-1\right) \cdot 3 - \left(3x+7\right) \cdot 2x}{\left(x^2-1\right)^2}$$

$$= \frac{3x^2 - 3 - 6x^2 - 14x}{x^4 - 2x^2 + 1}$$

$$= \frac{-3x^2 - 14x - 3}{x^4 - 2x^2 + 1}$$

You've Got Problems

Problem 3: Use the Quotient Rule to differentiate $f(x) = \dfrac{3x^4 + 2x^2 - 7x}{x-5}$ and simplify $f'(x)$.

The Chain Rule

This rule is used *a lot* in calculus. You need to use it whenever you are deriving a function that contains anything except a single variable, like x. For example, you can already find the derivative of $y = \sqrt{x}$, since you can rewrite it as $y = x^{1/2}$ and apply the Power Rule. The square root function contains only the x. However, without the Chain Rule, you cannot differentiate $y = \sqrt{2x}$, because there's something besides a plain old x inside the square root function. In other words, the Power Rule is great for vanilla functions, but now it's time for some Rocky Road.

The Chain Rule: The derivative of the composite function $h(x) = f(g(x))$ is $h'(x) = f'(g(x)) \cdot g'(x)$.

Critical Point

The derivatives of logarithmic and exponential equations use the Chain Rule heavily. Make sure to learn these patterns:

- $\frac{d}{dx}\left(\log_a f(x)\right) = \frac{1}{(\ln a)f(x)} \cdot f'(x)$

- $\frac{d}{dx}\left(a^{f(x)}\right) = (\ln a)a^{f(x)} \cdot f'(x)$

There are special cases for the natural logarithm ($\ln x$) and the natural exponential function (e^x), so you'll see those more often: $\frac{d}{dx}\left(\ln x\right) = \frac{1}{x}$ and $\frac{d}{dx}\left(e^x\right) = e^x$.

In other words, if one function contains another function, you must apply the Chain Rule to find the derivative. Here's how the process works: Take the derivative of the "outer" function, while leaving the "inner" function alone inside it. Now, multiply that by the derivative of the "inner" function.

Example 5: Use the Chain Rule to find the derivative of $y = \sqrt{3x + 1}$.

Solution: Remember, we have to use the Chain Rule because the outer function (the square root function) contains something other than a plain old x (it contains $3x + 1$). So, if we compare this example to the Chain Rule formula, the outer function is $f(x) = \sqrt{x}$ and the inner function is $g(x) = 3x + 1$. (Note that $f(g(x)) = \sqrt{3x + 1}$, our original function. We must always assign f and g so that we get that original function.) By the Power Rule, $f'(x) = \frac{1}{2\sqrt{x}}$, so $f'(g(x))$, which we need to complete the Chain Rule, is $\frac{1}{2\sqrt{3x+1}}$. The other piece we need, $g'(x)$, is very easy to find; it is 3 by the Power Rule. Now we can put it all together and get the following answer:

$$y' = f'(g(x)) \cdot g'(x)$$
$$= \frac{1}{2\sqrt{3x + 1}} \cdot 3$$
$$= \frac{3}{2\sqrt{2x + 1}}$$

You've Got Problems
Problem 4: Use the Chain Rule to differentiate $y = (x^2 + 1)^5$.

Rates of Change

Derivatives are so much more than what they seem. True, they give the slope of the tangent line to a curve. But that slope can tell us a great deal about the curve. One characteristic of the derivative we will exploit time and time again is this: *The derivative of a curve tells us the instantaneous rate of change of the curve.* This is key, because a curvy function changes at different rates throughout its domain—sometimes it's increasing quickly and the tangent line is steep (causing a high-valued derivative). At other places the curve may be increasing shallowly or even decreasing, causing the derivative to be small or negative, respectively. Look at the graph of $f(x)$ in Figure 9.2.

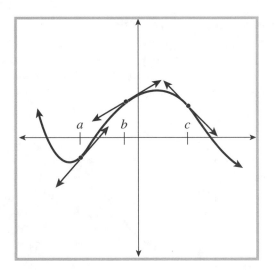

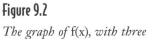
Figure 9.2

The graph of f(x), *with three points of interest shown.*

Critical Point

The graph of a line always changes at the exact same rate. For example, $g(x) = 4x - 3$ will always increase at a rate of 4, since that is the slope of the line and also the derivative. Since curves are not straight, they will not always possess the same slope as lines do. Therefore, we instead use the slope of a curve's *tangent* line to determine its rate of change. Because tangent lines vary depending on where on the curve they are drawn, a curve will possess different rates of change depending on where you look, as illustrated in Figure 9.2.

At $x = a$, f is increasing ever so slightly, causing the tangent line there to be shallow. Since a shallow line has a slope close to 0, the derivative here will be very small. In other words, the rate of change of the graph is very small at the instant that $x = a$. However, at the instant that $x = b$, the graph is climbing more rapidly, causing a steeper tangent line, which in turn causes a larger derivative. Finally, at $x = c$, the graph is decreasing at a pretty good rate, so the instantaneous rate of change there is negative (since the slope of the tangent line is negative).

Critical Point

Remember, the slope of a tangent line to a curve tells you the curve's rate of change at that value of x (i.e., the instantaneous rate of change, since you can only tell what's going on at that instant). The slope of the secant line to a curve tells you the *average* rate of change over the specified interval.

You can also use the slope of a *secant* line to determine rates of change on a graph. However, the slope of a secant line describes something different: the *average* rate of change over some portion of the graph. Finding the slope of a secant line is very easy, as you'll see in the next example.

Example 6: Poteet, Inc. has just introduced a new, revolutionary brand of athletic sock into the market. The new innovation is a special sweat-absorbing "cotton-esque" material that supposedly prevents foot odor. On their fourth day of sales, the snappy slogan, "If you smell feet, they ain't wrapped in Poteet's," was released, and sales immediately increased. Figure 9.3 is a graph of number of units sold versus the first six days of sales.

Figure 9.3

The rise of a new sweat-sock empire.

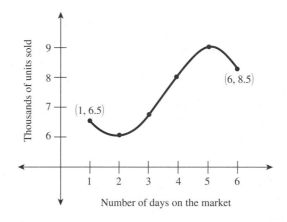

What was the average rate of units sold between day one and day six?

Solution: The problem asks us to find an average rate of change, which translates to finding the slope, m, of the secant line connecting the points $(1, 6.5)$ and $(6, 8.5)$. To do that, use our tried-and-true method from algebra:

$$m = \frac{y_2 - y_1}{x_2 - x_1}$$
$$= \frac{8.5 - 6.5}{6 - 1}$$
$$= \frac{2}{5}$$

That means Poteet's socks sold at a rate of two fifths of a thousand units per day, or $\frac{2}{5} \cdot 1{,}000 = 400$ units/day on average. So, even through the moderate decreases they experienced, the new slogan probably helped.

You've Got Problems

Problem 5: Given the function $g(x) = 3x^2 - 5x + 6$, find the following values:
  (a) the instantaneous rate of change of g when $x = 4$
  (b) the average rate of change on the x interval $[-1,3]$

Trigonometric Derivatives

Before we leave the land of simple derivatives, we must first discuss trigonometric derivatives. Each trig function has a unique derivative that you should automatically know. Whereas some are easy to build from scratch (as you'll see in Problem 6), others are quite difficult, so it's best to memorize the entire list. Trust me—a little memorization now goes a long way later. Take a deep breath and gaze upon the following list of important trig derivatives:

$$\frac{d}{dx}(\sin x) = \cos x \qquad \qquad \frac{d}{dx}(\arcsin x) = \frac{1}{\sqrt{1-x^2}}$$

$$\frac{d}{dx}(\cos x) = -\sin x \qquad \qquad \frac{d}{dx}(\arccos x) = -\frac{1}{\sqrt{1-x^2}}$$

$$\frac{d}{dx}(\tan x) = \sec^2 x \qquad \qquad \frac{d}{dx}(\arctan x) = \frac{1}{1+x^2}$$

$$\frac{d}{dx}(\cot x) = -\csc^2 x \qquad \qquad \frac{d}{dx}(\text{arccot } x) = -\frac{1}{1+x^2}$$

$$\frac{d}{dx}(\sec x) = \sec x \tan x \qquad \qquad \frac{d}{dx}(\text{arcsec } x) = \frac{1}{|x|\sqrt{x^2-1}}$$

$$\frac{d}{dx}(\csc x) = -\csc x \cot x \qquad \qquad \frac{d}{dx}(\text{arccsc } x) = -\frac{1}{|x|\sqrt{x^2-1}}$$

Critical Point

The notation $\frac{d}{dx}\big(f(x)\big)$ translates to "the derivative of the expression inside the parentheses." In other words, $\frac{d}{dx}\big(f(x)\big) = f'(x)$.

It's not as bad as you think—half of the inverse trig derivatives are different from the other half by only a negative sign.

You'll notice that I have included the inverse trig functions in this list, but you may not recognize them. Instead of using the notation $y = \sin^{-1}x$ to indicate the inverse sine, I use the notation $y = \arcsin x$. I am a huge fan of the latter notation, since $\sin^{-1}x$ looks a lot like $(\sin x)^{-1}$, which is equal to $\csc x$.

You'll have to be able to use these formulas with the Product, Quotient, and Chain Rules, so here are a couple of examples to get you used to them. Remember, if a trig function contains anything except a single variable x, you have to use the Chain Rule to find the derivative.

Example 7: If $f(x) = \cos x \sin 2x$, find $f'(x)$ and evaluate $f'\left(\frac{\pi}{2}\right)$.

Solution: Because this expression is the product of two variable expressions, you'll have to use the Product Rule. In addition, you'll have to use the Chain Rule to differentiate $\sin 2x$, since it contains more than just x inside the sine function. According to the Chain Rule, $\frac{d}{dx}\big(\sin 2x\big) = \cos(2x) \cdot 2 = 2\cos 2x$. Here's the Product Rule in action:

$$f'(x) = \cos x \cdot 2\cos 2x + \big(-\sin x\big)\big(\sin 2x\big)$$
$$f'\left(\frac{\pi}{2}\right) = 0 \cdot 2\cos\pi - 1 \cdot \sin\pi = 0 - 0 = 0$$

You've Got Problems

Problem 6: Use the Quotient Rule to prove that $\frac{d}{dx}\big(\cot x\big) = -\csc^2 x$.

The Least You Need to Know

- If a function is differentiable, it must also be continuous.
- A function is not differentiable at a point of discontinuity, a sharp point (cusp), or where the tangent line is vertical.
- The slope of a function's tangent line gives its instantaneous rate of change, and the slope of its secant line gives average rate of change.
- Simple derivatives (such as polynomials) can usually be found using the Power Rule.
- Products and quotients of variable expressions must be differentiated using the Product and Quotient Rules, respectively.
- You must use the Chain Rule to differentiate any function that contains something other than just x.

Chapter 10

Common Differentiation Tasks

In This Chapter

◆ Equations of tangent and normal lines
◆ Deriving equations with multiple variables
◆ Inverse derivatives
◆ Differentiating parametric equations

Even though the derivative is just the slope of the tangent line, its uses are innumerable. We've already seen that it describes the instantaneous rate of change of a nonlinear function. However, that hardly explains why it's one of the most revolutionary mathematical concepts in history. Soon we'll be exploring more (and substantially more exciting) uses for the derivative.

In the meantime, there's a little bit more grunt work to be done. (That makes you happy to read, doesn't it?) This chapter will help you perform specific tasks and find derivatives for very particular situations. Think of learning derivatives like trying to get your body in shape. Last chapter, you learned the basics, the equivalent of a good cardiovascular workout, working all of your muscles in harmony with each other. In this chapter, we're working out specific muscle groups, one section at a time. There's not a lot of similarity between each individual topic here, but exercising all of these abilities at the appropriate time (and knowing when that time arrives) is essential to being in shape mathematically.

Finding Equations of Tangent Lines

Writing tangent line equations is one of the most basic and foundational skills in calculus. You already know how to create the equation of a line using point-slope form (Chapter 2, "Polish Up Your Algebra Skills"). Since it's the equation of a tangent line you're after, the slope is the derivative of the function! All that's left to do is figure out the appropriate point, and if that were any easier, it'd be illegal.

Example 1: Write the equation of the tangent line to the curve $f(x) = 3x^2 - 4x + 1$ when $x = 2$.

Solution: Take a look at the graph of f in Figure 10.1 to get a sense of our task.

We want to find the equation of the tangent line to the graph at the indicated point (when $x = 2$). This is the point of tangency, where the tangent line will strike the graph. Therefore, this point is both on the curve *and* on the tangent line. Since point-slope form requires us to know a point on the line in order to create the equation of that line, we'll need to know the coordinates of this point. Since we already know the x-value, plug it into f to get the corresponding y-value:

$$f(2) = 3(2)^2 - 4 \cdot 2 + 1$$
$$= 12 - 8 + 1 = 5$$

Figure 10.1

The graph of
f(x) = 3x² − 4x + 1
and a future point
of tangency.

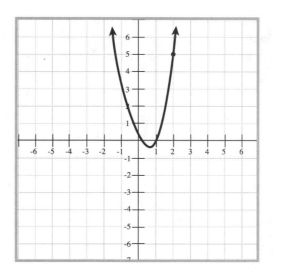

So, the point (2,5) is on the tangent line. Now all we need is the slope. By now you know that $f'(2)$ is, by definition, that exact slope, so let's find it, too:

$$f'(x) = 6x - 4$$
$$f'(2) = 6 \cdot 2 - 4 = 8$$

Now that we know a point on the tangent line and the correct slope, we'll slap those values into point-slope form and out pops the correct tangent line equation:

$$y - 5 = 8(x - 2)$$
$$y = 8x - 11$$

You've Got Problems

Problem 1: Find the equation of the tangent line to $g(x) = 3x^3 - x^2 + 4x - 2$ when $x = -1$.

Occasionally you'll be asked to find the equation of the *normal line* to a curve. Because the normal line is perpendicular to the tangent line at the point of tangency, you use the same point to create the normal line, but the slope of the normal line is the negative reciprocal of the slope of the tangent line. Back to Example 1 for a second. If we want to find the equation of the normal line to $f(x) = 3x^2 - 4x + 1$ when $x = 2$, we'd still use the point (2,5), but the slope would be $-\frac{1}{8}$ instead of 8. Once again, plugging these values into point slope form would complete the problem.

Talk the Talk

A **normal line** is perpendicular to a function's tangent line at the point of tangency.

Implicit Differentiation

I've mentioned the phrase "with respect to x" a few times, but now I need to describe to you exactly what that means. In 95 percent of your problems in calculus, the variables in your expression will match the variable you are "respecting" in that problem. For example, the derivative of $5x^3 + \sin x$, with respect to x, is $15x^2 + \cos x$. The fact that I said you were finding the derivative with respect to x didn't make the problem any harder or any different. In fact, I didn't have to tell you which variable you were "respecting," so to speak, because x was the only variable in the problem.

In this section, we'll take the derivative of functions containing x and y, and I will always ask you to find the derivative with respect to x. The only thing new to learn is this: What is the derivative of y with respect to x? The answer is this notation: $\frac{dy}{dx}$. It is literally read, "the derivative of y with respect to x." The numerator tells you what you're deriving, and the denominator tells you what you're respecting. One good thing is that you'll never have to find the derivative with respect to Rodney Dangerfield (since he never gets any respect).

Let's try a slightly more complex derivative. What is the derivative of $3y^2$, with respect to x? The first thing to notice is that the variable in the expression does not match the variable you're respecting, so you treat the y as a completely separate function and apply the Chain Rule. I know you're not used to using the Chain Rule when there's only a single variable inside the function, but if that variable is not the variable you're respecting, you have to give it a hard time and "rough it up" a little. So, to use the Chain Rule to derive $3y^2$, start by deriving the outer function and leaving y (the inner function) alone to get $6y$. Now you have to multiply this by the derivative of y with respect to x, and you get:

$$6y\frac{dy}{dx}$$

Critical Point

Whenever you take the derivative of a y expression with respect to x, it's basically the same thing as differentiating normally and then tacking on a $\frac{dy}{dx}$ at the end. For example, the derivative of $\tan x$ with respect to x is $\sec^2 x$. The derivative of $\tan y$ with respect to x is $\sec^2 y \cdot \frac{dy}{dx}$.

Talk the Talk

Implicit differentiation allows you to find the slope of a tangent line when the equation in question cannot be solved for y.

You will encounter differentiation such as this whenever you cannot solve the equation in question for y or $f(x)$. You may not have noticed, but every single derivative question until now has been worded "Find the derivative of y = ..." or "Find the derivative of $f(x)$ ...." When a problem asks you to find $\frac{dy}{dx}$ in an equation that cannot be solved for y, you have to resort to the process of *implicit differentiation*, which involves deriving variables with respect to other variables. Whereas in past problems the derivative would be indicated by y' or $g'(x)$, the derivative in implicit differentiation is indicated by $\frac{dy}{dx}$.

Example 2: Find the slope of the tangent line to the graph of $x^2 + 3xy - 2y^2 = -4$ at the point $(1,-1)$.

Solution: Yuck! Clearly this is not solved for y, and if you try to solve for y, you'll get discouraged quickly—mostly because solving it for y is impossible due to that blasted y^2. Implicit differentiation to the rescue. The first order of business is finding the derivative of each part of the equation with respect to x. Since you're new at this, we'll go term by term.

The derivative of x^2 with respect to x is $2x$. Nothing fancy is needed, since the variable in the term is the variable we're respecting. However, for the next term, $3xy$, we have to use the Product Rule, since there are two variable terms being multiplied ($3x$ and y). Remembering that the derivative of y, with respect to x, is $\frac{dy}{dx}$, the correct derivative of $3xy$ is $3x \cdot \frac{dy}{dx} + 3 \cdot y$. Finally, the derivative of $-2y^2$ is $-4y \cdot \frac{dy}{dx}$ and the derivative of -4 is 0.

Don't forget to differentiate on *both* sides of the equation! Even though I differentiate implicitly pretty often, I still sometimes forget to differentiate a constant term to get 0. I know; I am a lunkhead.

All together now, we get a derivative of:

$$2x + 3x\frac{dy}{dx} + 3y + 4y\frac{dy}{dx} = 0$$

Move all of the terms not containing a $\frac{dy}{dx}$ to the right side of the equation. Once you've done that, factor the common $\frac{dy}{dx}$ out of the terms on the left side of the equation:

$$3x\frac{dy}{dx} - 4y\frac{dy}{dx} = -2x - 3y$$

$$\frac{dy}{dx}(3x - 4y) = -2x - 3y$$

To finally get the derivative $\left(\frac{dy}{dx}\right)$ by itself, divide by $3x - 4y$ on both sides of the equation:

$$\frac{dy}{dx} = \frac{-2x - 3y}{3x - 4y}$$

That's the derivative. Now, the problem does ask us to evaluate it at $(1,-1)$, so plug those values in for x and y to get your final answer:

$$\frac{dy}{dx} = \frac{-2(1) - 3(-1)}{3(1) - 4(-1)} = \frac{-2 + 3}{3 + 4} = \frac{1}{7}$$

You've Got Problems

Problem 2: Find the slope of the tangent line to the graph of $4x + xy - 3y^2 = 6$ at the point $(3,2)$.

Differentiating an Inverse Function

Let's say you're given the function $f(x) = 7x - 5$ and asked to evaluate $\left(f^{-1}\right)'(1)$, the derivative of the inverse of f when $x = 1$. To find the answer, you would first find the inverse function (using the process we reviewed in Chapter 3) and then find the derivative.

However, did you know that you can evaluate the derivative of an inverse function *even if you can't find the inverse function itself?* (Insert dramatic soap opera music here.) You'll learn how to do it in just a second, but we have to review one skill first.

It's important that you're able to find values for an inverse function given only the original function before we try anything more difficult. The procedure we'll use is based on one of the most important properties of inverse functions: If the point (a,b) is on the graph of $f(x)$, then the point (b,a) is on the graph of $f^{-1}(x)$. In other words, if $f(a) = b$, then $f^{-1}(b) = a$.

Example 3: If $g(x) = x^3 + 2$, evaluate $g^{-1}(1)$.

Solution: *Method 1:* The easiest way to do this is to figure out exactly what $g^{-1}(x)$ is and then plug in 1. According to our procedure from Chapter 3, here's how you'd go about doing that:

$$y = x^3 + 2$$
$$x = y^3 + 2$$
$$y^3 = x - 2$$
$$g^{-1}(x) = \sqrt[3]{x - 2}$$

Therefore, $g^{-1}(1) = \sqrt[3]{1 - 2} = \sqrt[3]{-1} = -1$. However, there is another way to do this without actually finding $g^{-1}(x)$ first, and you need to learn that method.

Method 2: We're being asked the *output* of $g^{-1}(x)$ when its input is 1. Remember, I just said that $f(a) = b$ implies $f^{-1}(b) = a$, so therefore the output of g^{-1} when I input 1 is the *same exact thing* as the input of the original function g when I *output* 1. So, set the original function equal to 1 and solve; the solution will be $g^{-1}(1)$:

$$x^3 + 2 = 1$$
$$x^3 = -1$$
$$x = \sqrt[3]{-1} = -1$$

Critical Point

Here's a quick summary of this inverse function trick. If I want to evaluate $f^{-1}(a)$, a quick way to do it is to set $f(x) = a$ and solve.

Clearly, either method gives us the same answer.

Now that you possess this skill, we can graduate to finding values of the derivative of a function's inverse (say that 10 times fast, I dare you). As is the case with just about everything in calculus, there is a theorem governing this practice:

$$\left(f^{-1}\right)'(x) = \frac{1}{f'\left(f^{-1}(x)\right)}$$

You've Got Problems

Problem 3: Use the technique of Example 3, Method 2 to evaluate $f^{-1}(6)$ if $f(x) = \sqrt{2x^3 - 18}$.

So, evaluating the derivative is as simple as plugging the value into this slightly more complex, fractiony-looking guy. Once you substitute, your first objective will be to evaluate $f^{-1}(x)$ in the denominator (a skill which we just finished practicing, by no small coincidence).

Critical Point

Although this formula seems to appear out of thin air, it's pretty easy to create—here's how. Start with the simple inverse function property $f(f^{-1}(x)) = x$ and take the derivative with the Chain Rule:

$$f'\left(f^{-1}(x)\right) \cdot \left(f^{-1}\right)'(x) = 1$$

$$\left(f^{-1}\right)'(x) = \frac{1}{f'\left(f^{-1}(x)\right)}$$

Example 4: If $f(x) = x^3 + 4x + 1$, evaluate $\left(f^{-1}\right)'(2)$.

Solution: According to the formula you learned only moments ago:

$$\left(f^{-1}\right)'(2) = \frac{1}{f'\left(f^{-1}(2)\right)}$$

Start by evaluating $f^{-1}(2)$, which is the equivalent of solving the equation $x^3 + 4x + 1 = 2$. This is not an easy equation to solve; in fact, you can't do it by hand. You'll have to use some form of technology to solve the equation, whether it be a graphing calculator equation solver or a mathematical computer program. One way is to set the equation equal to $0(x^3 + 4x - 1 = 0)$ and calculate the x-intercept on a graphing calculator. Whichever method you choose, the answer is $x = .2462661722$, which you can plug into the formula:

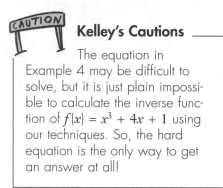

Kelley's Cautions

The equation in Example 4 may be difficult to solve, but it is just plain impossible to calculate the inverse function of $f(x) = x^3 + 4x + 1$ using our techniques. So, the hard equation is the only way to get an answer at all!

$$\left(f^{-1}\right)'(2) = \frac{1}{f'(.2462661722)}$$

$$= \frac{1}{3(.2462661722)^2 + 4}$$

$$= \frac{1}{4.1819410827}$$

$$\approx .239$$

I know that's a lot of decimals, but I didn't want to round any of them until the final answer, or it would have compounded the inaccuracy with every step.

You've Got Problems

Problem 4: If $g(x) = 3x^5 + 4x^3 + 2x + 1$, evaluate $\left(g^{-1}\right)'(-2)$.

Parametric Derivatives

In order to find a parametric derivative, you differentiate both the x and y components separately and divide the y derivative by the x derivative. In fancy-schmancy mathematical form, it looks like this:

$$\frac{dy}{dx} = \frac{\dfrac{dy}{dt}}{\dfrac{dx}{dt}}$$

This formula suggests that you should derive with respect to t, but you should derive with respect to whatever parameter appears in the problem. In the following example, for instance, you'll derive with respect to θ.

Example 5: Find the slope of the tangent line to the parametric curve defined by $x = \cos\theta$ and $y = 2\sin\theta$ when $\theta = \frac{5\pi}{6}$ (pictured in Figure 10.2).

Solution: Since the parameter in these equations is θ, the derivative of the set of parametric equations is given by …

$$\frac{dy}{dx} = \frac{\dfrac{dy}{d\theta}}{\dfrac{dx}{d\theta}}$$

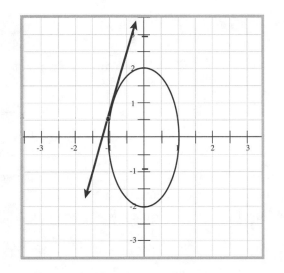

Figure 10.2

The graph of the parametric curve defined by x = cosθ *and* y = 2sinθ *with the tangent line drawn at* $\theta = \frac{5\pi}{6}$.

Simple differentiation tells you that $\frac{dx}{d\theta} = -\sin\theta$ and $\frac{dy}{d\theta} = 2\cos\theta$.

Finally, we need to calculate the derivative when $\theta = \frac{5\pi}{6}$:

$$\frac{dy}{dx} = \frac{2\cos\dfrac{5\pi}{6}}{-\sin\dfrac{5\pi}{6}} = \frac{2\left(-\dfrac{\sqrt{3}}{2}\right)}{-\left(\dfrac{1}{2}\right)} = 2\sqrt{3}$$

Kelley's Cautions

The second derivative (which, like all second derivatives, has the almost incomprhensible notation $\frac{d^2y}{dx^2}$) of parametric functions is *not* just the derivative of the first derivative. Instead, it is the derivative of the first derivative divided by the derivative of the *x* term again:

$$\frac{d^2y}{dx^2} = \frac{\dfrac{d}{dt}\left(\dfrac{dy}{dx}\right)}{\dfrac{dx}{dt}}$$

You've Got Problems

Problem 5: Evaluate $\frac{dy}{dx}$ and $\frac{d^2y}{dx^2}$ (the first and second derivatives) of the parametric equations $x = 2t - 3$ and $y = \tan t$.

The Least You Need to Know

◆ To write the equation of a tangent line, use the point of tangency and the derivative there in conjunction with point-slope form of a line.

◆ You must differentiate implicitly if the equation cannot be solved for y.

◆ The derivative of a function's inverse is given by the formula $\left(f^{-1}\right)'(x) = \frac{1}{f'\left(f^{-1}(x)\right)}$.

◆ To calculate a parametric derivative, divide the derivative of the y equation by the derivative of the x equation.

Using Derivatives to Graph

In This Chapter

- ◆ Critical points and relative extrema
- ◆ Understanding wiggle graphs
- ◆ Determining direction and concavity
- ◆ The Extreme Value Theorem

Though astrologers have maintained for decades that an individual's astrological sign provides insight into his or her personality, tendencies, and fate, many people remain unconvinced, deeming such thoughts absurd or (in extreme cases) poppycock. (This could be due to the fact that statements such as "The moon is in the third house of Pluto" sounds like the title of a new-age Disney movie.) Astrologers don't realize how close they actually came to the truth. It turns out that the signs of the derivatives of a function determine and explain the function's behavior.

In fact, the sign of the first derivative of a function explains what direction that function is heading, and the sign of the second derivative accurately predicts the concavity of the function. It is the third derivative of a function, however, that is able to predict when you will find true love, success in business, and how many times a week it's healthy to eat eggs for breakfast. The easiest way to visualize the signs of a function is via a wiggle graph, which sounds racy but is really quite ordinary when all is said and done.

Relative Extrema

One common human tendency is to compare oneself with his or her peers on a regular basis. You probably catch yourself doing this all the time, thinking things like, "Of all my friends, I am definitely the funniest." Perhaps you compare more mundane things, like being the best at badminton or having the loudest corduroy pants. However, when you go outside your social sphere, you often find someone who is significantly funnier than you or who possesses supersonically loud pants. This illustrates the difference between a relative extreme point and an absolute extreme point. You can be the smartest of a group of people without being the smartest person in the world.

For example, look at the graph in Figure 11.1, with points of interest *a*, *b*, *c*, *d*, and *e* noted.

Figure 11.1

This graph has only one absolute maximum and one absolute minimum, but several relative extrema.

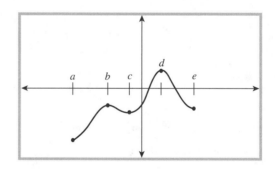

The absolute maximum on the graph occurs at *x = d*, and the absolute minimum of the graph occurs at *x = a*. However, the graph has *relative* maximums at *x = b* and *x = d*, and a relative minimum at *x = c*. These may not be the highest or lowest points of the entire graph, but (as little hills and valleys) are the highest and lowest points in their immediate vicinity (i.e., relative to that area).

Finding Critical Points

A critical number is an *x*-value where a graph could change direction (a *critical point* is the coordinate pair corresponding to that *x*-value). Mathematically, this happens whenever the function has a derivative of 0 or is undefined. Can you see the relationship between a critical point and a relative extreme point? If a graph hits a high or a low point, it will also change direction there. For example, if a graph has a relative maximum at *x = 1*, then the graph will be increasing just before that point and decreasing just after. Thus, all relative extrema must occur at critical points. This makes finding relative maximums and minimums very easy.

Talk the Talk

A **relative extreme** point (whether a maximum or a minimum) occurs when that point is higher or lower than all of the points in the immediate surrounding area. Visually, a relative maximum is the peak of a hill in the graph, and a relative minimum is the lowest point of a dip in the graph. Absolute extreme points are the highest or lowest of all the relative extrema on a graph. Remember that the term extrema is just plural for "extremely high or low point."

Example 1: Find the critical numbers of $f(x) = x^3 - x^2 - x + 2$.

Solution: Begin by finding the derivative of f; then, set it equal to 0 and solve:

$$f'(x) = 3x^2 - 2x - 1 = 0$$
$$(3x + 1)(x - 1) = 0$$
$$x = -\frac{1}{3}, 1$$

Talk the Talk

A **critical point** is a place where a graph *could* change direction, indicated mathematically by a 0 derivative or a point of nondifferentiability.

Since there are no places where $f'(x)$ does not exist, those are the only two critical points.

If you take a look at the graph of f, you'll notice that the graph does, indeed, change direction at those x-values (see Figure 11.2).

However, we don't need to use the graph of the function to determine (1) if the graph changes direction, or (2) if it does, whether it causes a relative maximum or minimum.

Classifying Extrema

As I alluded to earlier, the sign of the derivative tells you whether or not the graph is increasing or decreasing. This is true because an increasing graph will have a tangent line with a positive slope and a decreasing graph will possess a negatively sloped tangent line. Therefore, we can tell what's happening between the critical points (i.e., if the graph is increasing or decreasing) by picking some points between them and examining whether the derivative there is positive or negative.

Figure 11.2

The graph changes from increasing to decreasing at
$x = -\frac{1}{3}$, *and then returns to increasing once* x = 1.

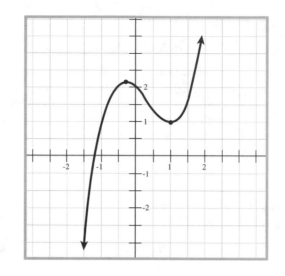

Example 2: Knowing that the critical numbers of $f(x) = x^3 - x^2 - x + 2$ are $x = -\frac{1}{3}, 1$ describe the direction of f between those critical numbers using the sign of $f'(x)$.

Solution: Choose three x-values, one before the first critical number, one between the critical numbers, and one after the second. I will choose simple values to make my life easier: $x = -1$, 0, and 2. Plug each of these x's into $f'(x)$, and the sign of the result will tell you if the function f is increasing or decreasing there:

$$f'(-1) = 3(-1)^2 - 2(-1) - 1 = 3 + 2 - 1 = 4$$
$$f'(0) = 3(0)^2 - 2(0) - 1 = -1$$
$$f'(2) = 3(2)^2 - 2(2) - 1 = 12 - 4 - 1 = 7$$

Because $f'(x)$ is positive when $x = -1$ and $x = 2$, f is increasing during those x-values. Since $x = -1$ comes before the first critical point and has a positive derivative, that means that the function will be increasing until $x = -\frac{1}{3}$, the first critical number. However, the derivative turns negative between the critical numbers, so f is decreasing between $x = -\frac{1}{3}$ and 1. After $x = 1$, the derivative turns positive again, so the function will increase beyond that point. All of this is summarized in the diagram in Figure 11.3.

You've Got Problems

Problem 1: Find the critical number for $h(x) = -x^2 + 6x + 27$ and determine whether or not it represents a relative maximum or minimum, based on the signs of $h'(x)$ before and after the critical number.

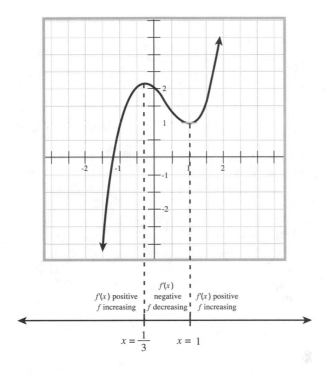

Figure 11.3

Notice how the sign of the derivative correlates with the direction of the original function.

$f'(x)$ positive
f increasing

$f'(x)$
negative
f decreasing

$f'(x)$ positive
f increasing

$x = \dfrac{1}{3}$ $x = 1$

The Wiggle Graph

A *wiggle graph* is a nice, compact way to visualize the signs of the derivative all at once. To create a wiggle graph, we'll use the procedure from Example 2. In other words, we'll find the critical numbers, pick "test points" between those critical numbers, and plug those into the derivative to determine the direction of the function. The result will be a number line, segmented by critical numbers, and labeled with the signs of the derivative for all of its intervals. This will help us to quickly find all relative extreme points on the graph.

Example 3: Create a wiggle graph for the function $f(x) = \dfrac{x^2 + 2x + 1}{x - 5}$ and use it to determine which critical numbers are relative extrema.

Solution: First we must find the critical numbers of f, that is, where $f'(x)$ is either equal to 0 or is undefined. We need the Quotient Rule to find $f'(x)$:

Talk the Talk

A **wiggle graph** (or sign graph) is a segmented number line that describes the direction of a function. It is created by finding critical numbers to determine interval boundaries, picking sample values from those intervals, and plugging the values into the derivative to obtain the proper sign. It's called a wiggle graph because it tells you which way the graph is wiggling (i.e., if it is increasing or decreasing).

$$f'(x) = \frac{(x-5)(2x+2)-(x^2+2x+1)}{(x-5)^2}$$

$$= \frac{x^2-10x-11}{(x-5)^2}$$

Since this is a fraction, it is equal to 0 when the numerator equals 0 and undefined when the denominator is equal to 0. Both of these events interest us for the purpose of finding critical numbers, so factor the numerator and set both it and the denominator equal to 0 and solve:

$$f'(x) = \frac{(x-11)\cdot(x+1)}{(x-5)^2}$$

The derivative equals 0 when $x = 11$ or -1 and is undefined when $x = 5$, so these are our critical numbers. Draw a number line and mark the numbers on it like Figure 11.4.

Figure 11.4

The beginnings of a wiggle graph.

This splits the number line into four intervals. Remember that the function will always go in the same direction during the entire interval, since it can only change direction at a critical number. Therefore, you can choose any number in each interval. To keep things simple, I'll choose the numbers $x = -2$, 0, 6, and 12. Now, find the derivative of f using the Quotient Rule and plug these three numbers into the derivative:

$$f'(-2) = \frac{(-13)(-1)}{(-7)^2} = \frac{13}{49}$$

$$f'(0) = \frac{(-11)(1)}{(-5)^2} = \frac{11}{25}$$

$$f'(6) = \frac{(-5)(7)}{1^2} = -35$$

$$f'(12) = \frac{(1)(13)}{7^2} = \frac{13}{49}$$

What matters are the signs of the derivatives, not so much the derivatives' values. Because $f'(-2)$ is positive, $f'(x)$ is actually positive for the entire interval $(-\infty,-1)$, so indicate that with a "+" sign above the interval in the wiggle graph. Do the same for the other intervals in Figure 11.5.

Now you can tell that the function only changes direction from increasing to decreasing at $x = -1$. If you plug that critical number into f, you get the critical point $(-1,0)$, which is a relative maximum. Similarly, a sign change at $x = 11$ indicates a relative minimum at the critical point $(11,24)$.

> **Critical Point** _____
>
> If a function changes from increasing to decreasing at a critical point, that point is a relative maximum. Similarly, a change from decreasing to increasing indicates a relative minimum.

$f'(x)$

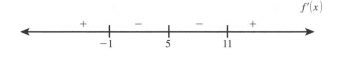

Figure 11.5

The signs of f'(x) correspond to the direction of f(x). Positive means increasing, negative means decreasing.

You've Got Problems

Problem 2: Draw the wiggle graph for the function $g(x) = 2x^3 - \frac{7}{2}x^2 - 3x + 10$, and explain on what intervals g is increasing.

The Extreme Value Theorem

Our first experience with existence theorems was the Intermediate Value Theorem. Do you remember it fondly? I guess that's a rhetorical question, because whether you liked it or not, here comes your second existence theorem. The Extreme Value Theorem, like its predecessor, really doesn't say anything earth-shattering, but it should make a lot of sense.

The Extreme Value Theorem: If a function $f(x)$ is continuous on the closed interval $[a,b]$, then f has an absolute maximum and an absolute minimum on $[a,b]$.

> **Kelley's Cautions** _____
>
> Before you conclude that a change of sign in a wiggle graph indicates a relative extreme point, make sure that the original function is defined there! For example, in the function $f(x) = \frac{1}{x^2}$, the function changes from increasing to decreasing around $x = 0$; verify with a wiggle graph of $f'(x)$. However, $x = 0$ is not in the domain of f, so it cannot be a relative maximum.

This theorem simply says that a piece of continuous function will always have a highest point and a lowest point. That's all. There's some place on that piece that's highest of all, and some place that's lowest, and these absolute extrema are guaranteed to be there (or your money back). Here's a little trick: A function's absolute extrema can only occur at one of two places—either at a relative extreme point or at an endpoint. This little trick makes finding the absolute extreme points very easy.

Example 4: Find the absolute maximum and absolute minimum for the function $f(x) = \frac{3}{5}x^5 - \frac{2}{3}x^3 - x + 2$ on the interval $[-2,1]$.

Solution: The absolute extrema we are to find are guaranteed to exist according to the Extreme Value Theorem, since f is a polynomial (and therefore continuous) on the closed interval. Start by drawing a wiggle graph for f, to determine where its relative extrema are. Same process as always: Set $f'(x) = 0$ and plug test points into the derivative:

$$f(x) = 3x^4 - 2x^2 - 1$$
$$= \left(3x^2 + 1\right)\left(x^2 - 1\right) = 0$$
$$x = 1, -1$$

Figure 11.6 represents the first derivative wiggle graph for $f(x)$. Because the sign of the derivative changes at both critical numbers (and they are both in the domain of f), we know that $x = -1$ and 1 both give rise to relative extrema and therefore possibly to absolute extrema as well.

Figure 11.6

According to this wiggle graph, f(x) *changes direction twice.*

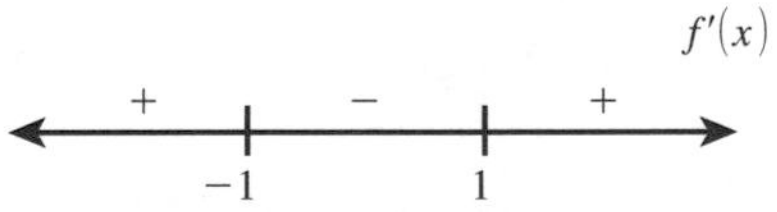

$f'(x)$

$$+ \qquad - \qquad +$$
$$-1 \qquad 1$$

CAUTION

Kelley's Cautions

There is no solution to the mini-equation $3x^2 + 1 = 0$ in Example 4, because solving it gives you $x = \pm\sqrt{-\frac{1}{3}}$, and you can't take the square root of a negative number.

Since an extreme value can only occur at a critical point ($x = -1$ or 1) or an endpoint ($x = -2$ or 1), plug all of those x-values into f to see which is the highest and which is the lowest:

$$f(-2) = -\frac{148}{15} \approx -9.867$$
$$f(-1) = \frac{46}{15} \approx 3.067$$
$$f(1) = \frac{14}{15} \approx .933$$

Therefore, the absolute maximum of f on the entire interval will be $\frac{46}{15}$ and the absolute minimum is $-\frac{148}{15}$. I know that those fractions were ugly, but whatever doesn't kill you makes you stronger, right? You're not buying that, are you?

Kelley's Cautions

In solving Example 4, a common error is reporting an absolute maximum of –1 and an absolute minimum of –2. Although these *are* the x-values where the extrema occur, they are not the extreme values themselves.

You've Got Problems

Problem 3: Find the absolute maximum and minimum for $g(x) = x^3 + 4x^2 + 5x - 2$ on the closed interval [–5,2].

Determining Concavity

Just as the first derivative describes the direction of the function, the sign of the second derivative describes the concavity of the original function. In other words, if $f''(x)$ is positive for some x-value, then $f(x)$ is concave up at that point. If, however, $f''(x)$ is negative, then $f(x)$ is concave down. What is concavity, though? Does it have anything to do with proper dental hygiene?

Critical Point

Not only does the sign of $f''(x)$ describe the concavity of $f(x)$, it also describes the direction of $f'(x)$. This is because $f''(x)$ actually is the first derivative of $f'(x)$, and remember that first derivatives describe the direction of their predecessors. For example, if $g''(2) = -7$ for some function g, then we know that $g(x)$ is concave down when $x = 2$ (since the second derivative is negative) *and* we know that $g'(x)$ is *decreasing* at $x = 2$.

Concavity describes the direction of a curve. A curve that is concave up is able to hold water poured into it. Let's break it down East Coast–style with a rhyme: If it's like a cup, then it's concave up.

Talk the Talk _____

The **concavity** of a curve describes the way the curve bends. Notice that the concave up in Figure 11.7 would catch water poured into it from above, whereas the concave down curve would deflect the water onto the floor, causing your mother to get angry.

Concavity describes how a curve bends. A curve that can hold water poured into it from the top of the graph is said to be concave up, whereas one that cannot hold water is said to be concave down. The sign of the second derivative reveals a function's concavity; if the function has a positive second derivative, then it is concave up. If, however, the second derivative of a function is negative, then it is concave down. You can remember this relationship between the second derivative's sign and concavity using Figure 11.8.

Figure 11.7

A tale of two curves whose second derivatives differ (you'll see what I mean soon).

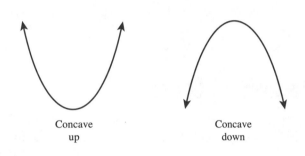

Concave
up

Concave
down

Figure 11.8

A smile is concave up, indicating a positive second derivative via the plus sign eyes. You'd be unhappy, too, if you were concave down.

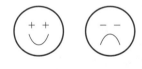

Talk the Talk _____

The point at which a graph changes concavity is called an **inflection point.**

Just like direction, however, the concavity of a curve can change throughout the function's domain (the points of change are called _inflection points_). We'll use a process that mirrors our first derivative wiggle graph to help us determine a function's concavity.

Another Wiggle Graph

Hopefully you've seen how useful a wiggle graph can be in visualizing a function's direction. It is just as useful in visualizing concavity, and is just as easy. This time, we'll use the second derivative to create the graph, and we'll plug test points into the second derivative rather than the first derivative to come up with the appropriate signs. Let's revisit an old friend, _f(x)_ from Example 4.

Example 5: On what intervals is the function $f(x) = \frac{3}{5}x^5 - \frac{2}{3}x^3 - x + 2$ concave up?

Solution: We'll create a wiggle graph with the second derivative. Notice that I label this wiggle graph $f''(x)$ (I labeled previous wiggle graphs $f'(x)$). This is responsible labeling, as it prevents confusion. Start by finding the second derivative:

$$f'(x) - 3x^4 - 2x^2 - 1$$
$$f''(x) = 12x^3 - 4x$$

Now, set that equal to 0 and solve for x to get your critical numbers:

$$4x(3x^2 - 1) = 0$$
$$x = 0, \pm\sqrt{\frac{1}{3}}$$

Don't forget the $\pm$ sign, because you are square-rooting both sides of an equation. It's time to draw the wiggle graph and choose test points just like before. Because you already know how to do this, let's jump straight to the correct graph in Figure 11.9.

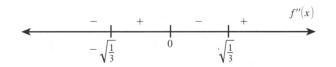

Figure 11.9

The second derivative wiggle graph for f(x).

We know f is concave up whenever $f''(x)$ is positive, so the correct intervals are $\left(-\sqrt{\frac{1}{3}}, 0\right)$ and $\left(\sqrt{\frac{1}{3}}, \infty\right)$.

You've Got Problems
Problem 4: When is the cosine function concave down on $(0, 2\pi)$?

The Second Derivative Test

Even though it is called the Second Derivative Test, its purpose is to tell whether or not an extreme point is a relative maximum or minimum (which we were determining using the signs of the *first derivative* and a wiggle graph earlier). The test uses the sign of the second derivative (and therefore the concavity of the graph at that point) to do all the work.

The Second Derivative Test: If you plug a critical number $x = a$ for the function $f(x)$ into $f''(x)$ and the result is positive, that critical number gives rise to a relative minimum on $f(x)$. If the result is negative, $x = a$ represents a relative maximum on $f(x)$. If the result is 0, you cannot draw any conclusion from the Second Derivative Test and must resort to the first derivative wiggle graph.

Example 6: Classify all the relative extrema of the function $g(x) = 3x^3 - 18x + 1$ using the Second Derivative Test.

Solution: First find the relative extrema of g like we did earlier in the chapter:

$$g'(x) = 9x^2 - 18 = 0$$
$$9x^2 = 18$$
$$x^2 = 2$$
$$x = \pm\sqrt{2}$$

Critical Point

If you think about it, the only possible extreme point you can have on a concave-up graph is a relative minimum—consider the point $x = 0$ on the graph of $y = x^2$ as an example.

Now plug both $\sqrt{2}$ and $-\sqrt{2}$ into $g''(x) = 18x$. Since $g''\left(\sqrt{2}\right) = 18\sqrt{2}$, which is positive, $x = \sqrt{2}$ represents the location of a relative minimum (according to the Second Derivative Test) and, conversely, since $g''\left(-\sqrt{2}\right) = -18\sqrt{2}$, that represents a relative maximum.

The Least You Need to Know

- Critical numbers are x-values that cause the derivative of a function to be 0 or undefined.
- If $f'(x)$ is positive, then $f(x)$ is increasing; a negative $f'(x)$ indicates a decreasing $f(x)$.
- If $f''(x)$ is positive, then $f(x)$ is concave up; a negative $f(x)$ indicates a concave-down $f(x)$.
- The first derivative wiggle graph and the Second Derivative Test are both techniques used for classifying relative extreme points.

Derivatives and Motion

In This Chapter

- ◆ What is a position equation?
- ◆ The relationship of position, velocity, and acceleration
- ◆ Speed vs. velocity
- ◆ Understanding projectile motion

Mathematics can actually be applied in the real world. This may shock and appall you, but it's very true. It's probably shocking because most of the problems we've dealt with have been purely computational in nature, completely devoid of correlation to real life. (For example, estimating gas mileage is a useful mathematical real-life skill, whereas factoring difference of perfect cube polynomials is not as useful in a straightforward way.) Most people hate real-life application problems because they are (insert scary wolf howl here) *word problems!*

Factoring and equation solving may be rote, repetitive, and a little boring, but at least they're predictable. How many nights have you gone to sleep haunted by problems like, "If Train *A* is going from Pittsburgh to Los Angeles at a rate of 110 kilometers per hour and Train *B* is traveling 30 kilometers less than half the number of male passengers in Train *A*, and the heading of Train *B* is 3 degrees less than the difference of the prices of a club sandwich on each train, then at what time will the conductor of the first train remember that he forgot to set his VCR to tape *Jeopardy?*"

Position equations are a nice transition into calculus word problems. Even though they are slightly bizarre, they follow clear patterns. Furthermore, they give you the chance to show off your new derivative skills.

The Position Equation

A *position equation* is an equation that mathematically models something in real life. Specifically, it gives the position of an object relative to something at a specified time.

Talk the Talk

A **position equation** is a mathematical model that outputs an object's position at a given time t. Position is usually given with relation to some fixed landmark, like the ground or the origin, so that a negative position means something. For example, $s(5) = -6$ may mean that the object in question is 6 feet below the origin after 5 seconds have passed.

Different books and teachers use different notation, but I will always indicate a position equation with the notation $s(t)$ for consistency. By plugging values of t into the equation, you can determine where the object in question was at that moment in time. Just in case you're starting to get stressed out, I'll insert something cute and cuddly into the mix—my kitten, Peanut.

Peanut pretty much has the run of the basement, and her favorite pastime (apart from her strange habit of chewing on my eyeglasses) is batting a ball back and forth along one of the basement walls. For the sake of ease, let's say the wall in question is 20 feet long; we'll call the exact middle of the wall position 0, the left edge of the wall position –10 and the right edge of the wall position 10, as in Figure 12.1.

Figure 12.1

The domain of Peanut the cat. For the sake of reference, I have labeled the middle and edges of the room.

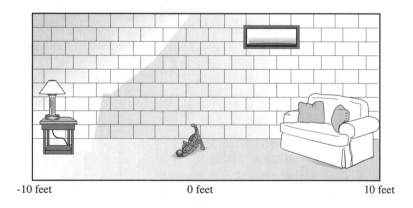

-10 feet   0 feet   10 feet

Let's examine the kitten's position versus time in a simple example. We'll keep returning to this example throughout the chapter as we compound our knowledge of derivatives and motion.

Example 1: During the first four seconds of a particularly frisky playtime, Peanut's position (in feet at time = t seconds) along the wall is given by the equation $s(t) = t^3 - 3t^2 - 2t + 1$. Evaluate and explain what is meant by $s(0)$, $s(2)$, and $s(4)$.

Solution: Plug each number into s. A positive answer means she is toward the right of the room (our right as we face the wall), whereas a negative answer means she is to the left of center. The larger the number, the farther she is to the right or left:

$$s(0) = 0^3 - 3 \cdot 0^2 - 2 \cdot 0 + 1 = 1$$
$$s(2) = 2^3 - 3 \cdot 2^2 - 2 \cdot 2 + 1 = 8 - 12 - 4 + 1 = -7$$
$$s(4) = 4^3 - 3 \cdot 4^2 - 2 \cdot 4 + 1 = 64 - 48 - 8 + 1 = 9$$

Therefore, when $t = 0$ (i.e., before we start measuring elapsed time), $s(0) = 1$ tells us that cat began 1 foot to the right of the center. Two seconds later ($t = 2$), she had used her lightning-fast kitty movements to travel 8 feet to our left, meaning she was then only 3 feet from the left wall. Two seconds after that ($t = 4$) she had moved 16 feet to our right, only 1 foot away from the right-hand wall. That is one fast-moving cat, for Pete's sake!

There's nothing really fancy about the position equation; given a time input, it basically tells you where the object in question was at that time. Notice, though, that the position equation in Example 1 is a nice, continuous, and differentiable polynomial. We can find the derivative awfully easily, but what does the derivative of the position equation represent?

Critical Point

The value $s(0)$ is often called initial position, since it gives the position of the object before you start measuring time. Similarly, $v(0)$ and $a(0)$ are the initial velocity and initial acceleration.

You've Got Problems

Problem 1: A particle moves vertically (in inches) along the y-axis according to the position equation $s(t) = \frac{1}{2}t^3 - 5t^2 + 3t + 6$, where t represents seconds. At what time(s) is the particle 30 inches below the origin?

Velocity

Remember that the derivative describes the rate of change of a function. Therefore, $s'(t)$ describes the velocity of the object in question at any given instant. It makes sense that velocity is equivalent to the rate of change of position, since velocity measures how quickly it takes you to get from one position to another. Many people are confused about the difference between speed and velocity, since both measure the rate at which something travels.

The official definition of velocity is that it compounds an object's speed with the component of direction, whereas speed just gives you the rate at which the object is traveling. Practically speaking, this means that velocity can be negative, but speed cannot. What does a negative velocity mean? It depends on the problem. In a horizontal motion problem (like the Peanut the cat problem), it means velocity towards the left (since the left was defined as the negative direction). In a vertical motion problem, a negative velocity typically means that the object is heading downwards.

Critical Point

If an object is moving downward at a rate of 15 feet per second, you could say that its velocity is −15 ft/sec, whereas its speed is 15 ft/sec. *Speed is always the absolute value of velocity.*

Therefore, to find the velocity of an object at any instant, use the derivative and plug in the desired time for t. If, however, you want an object's average velocity (i.e., average rate of change), remember that this value comes from the slope of the secant line. Remember how quickly Peanut was darting around in Example 1? Let's get those exact speeds using the derivative.

Example 2: Peanut the cat's position, in feet, for $0 \leq t \leq 4$ seconds is given by $s(t) = t^3 - 3t^2 - 2t + 1$. Find her velocity and speed at times $t = 1$ and $t = 3.5$ seconds, and give her average velocity over the t interval [1,3.5].

Solution: There are lots of parts to this problem, but none are hard. Let's start by calculating her velocity at the given times. Remember that the velocity is the first derivative of the position equation, so $s'(t) = v(t) = 3t^2 - 6t - 2$:

$$v(1) = 3 - 6 - 2 = -5 \text{ ft/sec}$$
$$v(3.5) = 36.75 - 21 - 2 = 13.75 \text{ ft/sec}$$

Peanut is moving at a speed of 5 ft/sec to the left (since the velocity is negative) at $t = 1$ second, and she is moving much faster, at a speed of 13.75 ft/sec, to the right, when $t = 3.5$ seconds. Now, let's find the average velocity. To do this, we'll need the coordinates of the two points on the position graph marking the endpoints of the interval. Plug the t values into s:

$$s(1) = 1 - 3 - 2 + 1 = -3$$
$$s(3.5) = 42.875 - 36.75 - 7 + 1 = .125$$

Now calculate the secant slope using the points $(1,-3)$ and $(3.5,.125)$:

$$m = \frac{.125 - (-3)}{3.5 - 1} = 1.25 \text{ ft/sec}$$

Therefore, even though she runs left and right at varying speeds, she averages a speed of 1.25 ft/sec towards the right over the time interval [1,3.5].

Kelley's Cautions

Remember, the tangent line to position gives you instantaneous velocity, and the secant line to position gives you average velocity. If you do the secant line to velocity, you get something completely different, so make sure you plug $t = 1$ and

---

You've Got Problems

Problem 2: A particle moves vertically (in inches) along the y-axis according to the position equation $s(t) = \frac{1}{2}t^3 - 5t^2 + 3t + 6$, where t represents seconds. Rank the following from least to greatest: the speed when $t = 3$, the velocity when $t = 7$, and the average velocity on the interval [2,6].

Acceleration

As velocity is to position, so is acceleration to velocity. In other words, acceleration is the rate of change of velocity. Think about it—if you're driving in a car whose speed suddenly changes, that sense of being pushed back in your seat is due to the effects of acceleration. It is not the high rate of speed that makes roller coasters so scary. Aside from their height, it is the sudden acceleration and deceleration of the rides that causes the passengers to experience dizzying effects (and occasionally their previous meal).

To calculate the acceleration of an object, evaluate the second derivative of position (or the first derivative of velocity). To calculate average acceleration, find the slope of the secant line on the velocity function (for the same reasons that average velocity is the secant slope on the position function). Let's head back to the cat of mathematical mysteries one last time.

Critical Point

The units for acceleration will be the same as the units for velocity, except the denominator will be squared. For example, if velocity is measured in feet per second (ft/sec), then acceleration is measured in feet per second per second, or ft/sec².

Critical Point _____

If the first derivative of position represents velocity and the second derivative represents acceleration, the third derivative represents "jerk," the rate of change of acceleration. Think of jerk as that feeling you get as you switch gears in your car and the acceleration changes. I've never seen a problem concerning jerk, but I have known a few mathematicians who were pretty jerky.

Example 3: Peanut the cat's position, in feet, at any time $0 \leq t \leq 4$ seconds is given by $s(t) = t^3 - 3t^2 - 2t + 1$. When, on the interval [0,10], is she decelerating?

Solution: Since the sign of the second derivative determines acceleration, we want to know when $s''(x)$ is negative. So, make a $s''(x)$ wiggle graph by setting it equal to 0, finding critical numbers, and picking test points (as we've done in the past). The wiggle graph for the second derivative is given in Figure 12.2.

$$s'(t) = v(t) = 3t^2 - 6t - 2$$
$$s''(t) = a(t) = 6t - 6$$
$$6t - 6 = 0$$
$$t = 1$$

Figure 12.2

Since s''(x) is negative on (0,1), she is decelerating on that interval.

You've Got Problems

Problem 3: A particle moves vertically (in inches) along the y-axis according to the position equation $s(t) = \frac{1}{2}t^3 - 5t^2 + 3t + 6$, where t represents seconds. At what time t is the acceleration of the particle equal to -1 in/sec²?

Projectile Motion

One of the easiest types of motion to model in elementary calculus is projectile motion, the motion of an object being acted upon solely by gravity. Did you ever notice that anytime you throw something, it follows a parabolic path to the ground? It is very easy to write the position equation describing that path with only a tiny bit of information. Mind you, these equations can't give you the _exact_ position, since ignoring wind resistance and drag makes the problem much easier.

Critical Point _____

Unquestionably, one of the grossest examples of projectile motion is in the movie _The Exorcist_. We will not be doing any examples involving pea soup.

Scientists often pooh-pooh these little pseudoscientific math applications, saying that ignoring such factors as wind resistance and drag renders these examples worthless. Math people usually contend that, although not perfect, these examples show how useful even a simple mathematical concept can be.

The position equation of a projectile looks like this:

$$s(t) = -\tfrac{1}{2}g \cdot t^2 + v_0 \cdot t + h_0$$

You plug the object's initial velocity into v_0, the initial height into h_0, and the appropriate gravitational constant into g (which stands for acceleration due to gravity)—if you are working in feet, use $g = 32$, whereas $g = 9.8$ if the problem contains meters. Once you create your position equation, by plugging into the formula, it will work just like the other position equations from this chapter, outputting the vertical height of the object in relation to the ground (for example, a position of 12 translates to a position of 12 feet above ground).

Example 4: Here's a throwback to 1970s television for you. A radio station called WKRP in Cincinnati is running a radio promotion. For Thanksgiving, they are dropping turkeys from the station's traffic helicopter into the city below, but little do they know that turkeys are not so good at the whole flying thing. Assuming that they were tossed with a miniscule initial velocity of 2 ft/sec from a safe hovering height of 1,000 feet above ground, how long does it take a turkey to hit the road below, and at what speed will the turkey be traveling at that time?

Critical Point

Notice that $s''(t)$ always equals either -32 or -9.8, depending on the units of the problem. This is because the only acceleration acting on the object is gravity.

Solution: The problem contains feet, so we use $g = 32$ ft/sec^2; we're also given $v_0 = 2$ and $h_0 = 1,000$, so plug these into the formula to get the position equation of $s(t) = -16t^2 + 2t + 1,000$. We want to know when they hit the ground, which means they have a position of 0, so solve the equation $-16t^2 + 2t + 1,000 = 0$. You can use a calculator or the quadratic formula to come up with the answer of $t = 7.9684412$ seconds (the other answer of -7.84 doesn't make sense—a negative answer suggests going back in time, and that's never a good idea, especially with poultry). If you plug that value into $s'(x)$, you'll find that the turkeys were falling at a velocity of -252.990 ft/sec. Oh, the humanity.

You've Got Problems

Problem 4: If a cannonball is fired from a hillside 75 meters above ground with an initial velocity of 100 meters/second, what is the greatest height the cannonball will reach?

The Least You Need to Know

◆ The position equation tells you where an object is at any time t.

◆ The derivative of the position equation is the velocity equation, and the derivative of velocity is acceleration.

◆ If you plug 0 into position, velocity, or acceleration, you'll get the initial value for that function.

◆ The formula for the position of a projectile is $s(t) = -\frac{1}{2} g \cdot t^2 + v_0 \cdot t + h_0$.

Common Derivative Applications

In This Chapter

◆ Limits of indeterminate expressions

◆ The Mean Value Theorem

◆ Rolle's Theorem

◆ Calculating related rates

◆ Maximizing and minimizing functions

It's been a fun ride, but our time with the derivative is almost through. Don't get too emotional yet—I've saved the best for last, and this chapter will be a hoot (if you like those dreaded word problems, that is). Just like in the last chapter, we'll be looking at the relationship between calculus and the real world, and you'll probably be surprised just what you can do with very simple calculus procedures.

This chapter has it all: cool shortcuts, a few more existence theorems, romance, adventure, and the two topics most first-year calculus students find the trickiest. For the sake of predictability, we'll go through the topics in the order of difficulty, starting with the easiest and progressing to the more advanced.

Evaluating Limits: L'Hôpital's Rule

Way, way back, many chapters ago, in a galaxy far, far away, you were stressed about limits. Since then, you've had a whole lot more to stress about, so it's high time we destressed you a bit. Little did you know that as you were plugging away, learning derivatives, you also learned a terrific shortcut for finding limits. This shortcut (called *L'Hôpital's Rule*) can be used to find limits that, after substitution, are in indeterminate form.

Critical Point

L'Hôpital's Rule can only be used to calculate limits that are indeterminate (i.e., the value cannot immediately be found). The most common indeterminate forms are $\frac{\pm\infty}{\pm\infty}, \frac{0}{0}, \frac{0}{0}$, and $0 \cdot \infty$.

To show just how useful L'Hôpital's Rule is, we'll return briefly to Chapter 6 and fish out two limits we couldn't previously calculate by hand. These two limits will comprise the next ex-ample. The first limit we could only memorize (but couldn't justify via any of our methods at the time). We calculated the second limit using a little trick (comparing degrees for limits at infinity), but that method was a trick only. We had no proof or justification for it at all. Finally, a little pay dirt for the curious at heart.

L'Hôpital's Rule: If $h(x) = \frac{f(x)}{g(x)}$ and $\lim\limits_{x \to c} h(x)$ is in indeterminate form (e.g., $\frac{0}{0}$ or $\frac{\infty}{\infty}$),

then $\lim\limits_{x \to c} h(x) = \lim\limits_{x \to c} \frac{f'(x)}{g'(x)}$. In other words, take the derivative of the numerator and denominator separately (not via the Quotient Rule) and substitute in c again to find the limit.

Kelley's Cautions

You can only use L'Hôpital's Rule (pronounced *Low-PEE-towels*) if you have indeterminate form after substituting—it will not work for other, more common, limits.

Example 1: Calculate both of the following limits using L'Hôpital's Rule:

a) $\lim\limits_{x \to 0} \frac{\sin x}{x}$

Solution: If you substitute in $x = 0$, you get $\frac{\sin 0}{0} = \frac{0}{0}$,

and $\frac{0}{0}$ is in indeterminate form. So, apply L'Hôpital's Rule by taking the derivative of sin x (which is cos x) and the derivative of x (which is 1) and replacing those pieces with their derivatives:

$$\lim\limits_{x \to 0} \frac{\cos x}{1}$$

Now, substituting won't give you $\frac{0}{0}$. In fact, substituting gives you cos 0, which equals 1. We memorized that 1 was the answer in Chapter 6, but now we know why.

$$\text{b) } \lim_{x \to \infty} \frac{5x^3 + 4x^2 - 7x + 4}{2 + x - 6x^2 + 8x^3}$$

Solution: If you plug in $x = \infty$, for all the x's you get a huge number on top divided by a huge number on the bottom $\left(\frac{\infty}{\infty}\right)$, which is indeterminate form, so apply L'Hôpital's Rule:

$$\lim_{x \to \infty} \frac{15x^2 + 8x - 7}{1 - 12x + 24x^2}$$

Uh oh. Substitution *still* gives you $\frac{\infty}{\infty}$. Never fear—keep applying L'Hôpital's Rule until it gives you a legitimate answer:

$$\lim_{x \to \infty} \frac{30x + 8}{-12x + 48x}$$

$$\lim_{x \to \infty} \frac{30}{48} = \frac{30}{48} = \frac{5}{8}$$

Once there are no more x's in the problem $\left(\lim_{x \to \infty} \frac{30}{48}\right)$, no substitution is necessary, and the answer falls out like a ripe fruit.

You've Got Problems

Problem 1: Evaluate $\lim_{x \to \infty} \left(x^{-2} \cdot \ln x\right)$ using L'Hôpital's Rule. *Hint:* Begin by writing the expression as a fraction.

More Existence Theorems

Man has struggled over what exactly determines life and what defines existence. Descartes once mused, "I think, therefore I am," suggesting that thought defined existence. Most calculus students go one step further, lamenting, "I am in mental anguish, therefore I am in calculus." Philosophy aside, the next two theorems don't try to answer such deep questions; they simply state that something exists, and that's good enough for them.

The Mean Value Theorem

This neat little theorem gives an explicit relationship between the average rate of change of a function (i.e., the slope of the secant line) and the instantaneous rate of change of a function (i.e., the slope of the tangent line). Specifically, it guarantees that at some point on a closed interval, the tangent line will be parallel to the secant line for that interval (see Figure 13.1).

Figure 13.1

Here, the secant line is drawn connecting the endpoints of the closed interval [a,b]. At x = c, which is on that interval, the tangent line is parallel to the secant line.

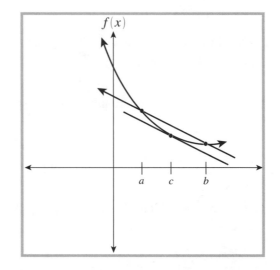

Mathematically, parallel lines have equal slopes. Therefore, there is always some place on an interval where a continuous function is changing at exactly the same rate it's changing on average for the entire interval. Now that you have an idea of what it means, here is what the theorem looks like:

The Mean Value Theorem: If a function $f(x)$ is continuous and differentiable on the closed interval $[a,b]$, then there exists a point c between a and b such that $f'(c) = \frac{f(b)-f(a)}{b-a}$. In other words, a point c is guaranteed to exist such that the derivative there ($f'(c)$) is equal to the slope of the secant line for the interval $\left(\frac{f(b)-f(a)}{b-a} \right)$.

Example 2: At what x-value(s) on the interval $[-2,3]$ does the graph of $f(x) = x^2 + 2x - 1$ satisfy the Mean Value Theorem?

Solution: Somewhere, the derivative must equal the secant slope; so, let's start by finding the derivative:

$$f'(x) = 2x + 2$$

That was easy. Now on to the secant slope. To calculate it, plug –2 and 3 into the function to get the secant's endpoints: (–2,–1) and (3,14). Now, find the secant slope:

$$\frac{14 - (-1)}{3 - (-2)} = \frac{15}{5} = 3$$

Therefore, at some point on the interval, the derivative and 3 are guaranteed to equal each other:

$$2x + 2 = 3$$
$$2x = 1$$
$$x = \frac{1}{2}$$

If you look at the graph of f in Figure 13.2, it does, indeed, appear that the tangent line at $x = \frac{1}{2}$ is parallel to the secant line connecting (–2,–1) and (3,14).

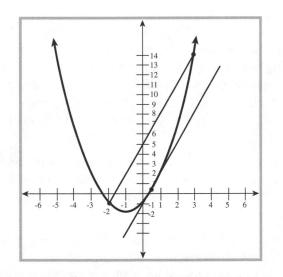

Figure 13.2

Equal secant and tangent slopes mean parallel lines.

You've Got Problems

Problem 2: Given the function $g(x) = \frac{1}{x}$, find the x-value(s) that satisfies the Mean Value Theorem on the interval $\left[\frac{1}{4}, 1\right]$.

Rolle's Theorem

Rolle's Theorem is a specific case of the Mean Value Theorem. It says that if the slope of the secant line on an interval is 0 (in other words, the secant line is horizontal) then somewhere on that interval, the tangent slope will also be 0. Since you already understand the Mean Value Theorem, this isn't new information. Our previous theorem guaranteed the lines would have the same slope no matter what the secant slope was. Here's what Rolle's Theorem looks like mathematically:

Rolle's Theorem: If a function $f(x)$ is continuous and differentiable on a closed interval $[a,b]$ and $f(a) = f(b)$, then there exists a c between a and b such that $f'(c) = 0$.

Let's prove this with the Mean Value Theorem—it guarantees that the secant slope will equal the tangent slope somewhere on $[a,b]$. The secant slope connecting the points $(a,f(a))$ and $(b,f(b))$ is $\dfrac{f(b)-f(a)}{b-a}$, but since the theorem states that $f(a) = f(b)$, this fraction becomes $\dfrac{0}{b-a} = 0$. Therefore, the slope of the secant line is 0. According to the Mean Value Theorem, $f'(x)$ has to equal 0 somewhere inside the interval, at a point Rolle's Theorem calls c.

Related Rates

Related rates problems are among the most popular problems (for teachers) and feared problems (for students) in calculus. You can tell if a given problem is a related rates problem because it will contain wording like "how quickly is … changing?" Related rates problems help you figure out how quickly one variable in a problem is changing if you know how quickly another variable is changing. No two problems will be alike, but the procedure is exactly the same for all problems of this type, and they actually get to be sort of fun once you get used to them.

Let's walk through a classic related rates problem: the ladder sliding down the side of a house dilemma. The only step that will differ between this and any other problem is the very first: finding an equation that characterizes the situation. Once you get past that initial step, everything is smooth sailing.

Example 3: Goofus and Gallant (you might remember them from the *Highlights* magazine), are painting my house. Whereas Gallant properly secured his 13-foot ladder before climbing it, Goofus does not, and as he climbs his ladder, it slides down the side of the house at a constant rate of 2 feet/second. How quickly is the base of the ladder sliding horizontally away from the house when the top of the ladder is 5 feet from the ground?

Solution: You can tell it's a related rates problem because it's asking you to find how quickly something is changing or moving. I always start these by drawing a picture of the situation (see Figure 13.3).

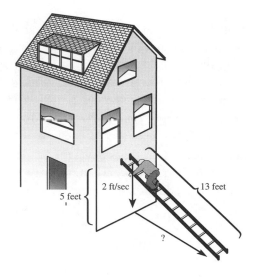

Figure 13.3

Recipe for disaster: the 13-foot ladder, with its top only 5 feet from the ground, and Goofus heroically clinging to it.

We need to pick an equation that represents the situation. Notice that the ladder, the house, and the ground make a right triangle; the problem gives me information about the lengths of the legs of a right triangle. Therefore, I will use the Pythagorean Theorem as my primary equation, as it relates the lengths of the sides of a right triangle. To make it easier to visualize, I will strip away all of the extraneous visual information:

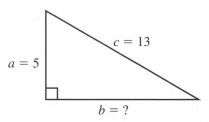

Figure 13.4

Goofus's predicament, minus the clever illustrations.

According to the picture I've drawn in Figure 13.4 (and the Pythagorean theorem), I know that $a^2 + b^2 = c^2$. Warning: Don't plug in any values you know (like $a = 5$) until we complete the next step, which is differentiating everything with respect to t:

$$2a\frac{da}{dt} + 2b\frac{db}{dt} = 2c\frac{dc}{dt}$$

Kelley's Cautions

Remember, you won't use the Pythagorean theorem for every related rates problem. You'll have to pick your primary equation based on the situation. Look at Problem 3 for a different example.

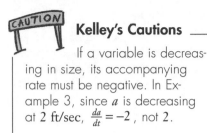

Kelley's Cautions

If a variable is decreasing in size, its accompanying rate must be negative. In Example 3, since a is decreasing at 2 ft/sec, $\frac{da}{dt} = -2$, not 2.

You might be wondering, "What does $\frac{da}{dt}$ mean?" It represents how quickly a is changing. The problem tells us that the ladder is falling, so side a is actually getting smaller at a rate of 2 ft/sec, so we write $\frac{da}{dt} = -2$. We have no idea what $\frac{db}{dt}$ equals, because that's the quantity we are looking for. However, we do know that $\frac{dc}{dt} = 0$, because c (the length of the ladder) will not change as it slides down the house. Now, we know most of the variables in the equation. In fact, you can even find that $b = 12$ using the Pythagorean Theorem, knowing that the other sides of the triangle are 5 and 13. So, let's plug in everything we know:

$$2 \cdot 5 \cdot (-2) + 2 \cdot 12 \cdot \frac{db}{dt} = 2 \cdot 13 \cdot 0$$

All you have to do is solve for $\frac{db}{dt}$ and you're finished:

$$-20 + 24\frac{db}{dt} = 0$$

$$\frac{db}{dt} = \frac{20}{24} = \frac{5}{6} \text{ ft/sec}$$

Therefore, b is increasing at a rate of $\frac{5}{6}$ ft/sec, so that's how quickly the base of the ladder is sliding away from the house.

Here are the steps to completing a related rates problem:

1. Construct an equation containing all the necessary variables.
2. Before substituting any values, derive the entire equation with respect to t.
3. Plug in values for all the variables except the one for which you're solving.
4. Solve for the unknown variable.

You've Got Problems

Problem 3: You've heard it's a bad idea to buy pets at mall pet stores, but you couldn't resist buying an adorable little baby cube. Well, after three months of steady eating, it's begun to grow. In fact, its volume is increasing at a constant rate of 5 cubic inches a week. How quickly is its surface area increasing when one of its sides measures 7 inches?

Optimization

Even though this is the most feared of all differentiation applications, I have never understood why. When you're looking for the biggest or smallest something can get (i.e., optimizing), all you have to do is create a formula representing that quantity and then find the relative extrema using wiggle graphs. You've been doing these all along; don't get freaked out unnecessarily. To explore optimization, we'll again examine a classic calculus problem that has haunted students like you for years and years.

Example 4: What are the dimensions of the box containing the largest volume that can be made by cutting congruent squares from the corners of a piece of paper measuring 11 by 14 inches? (Assume that the box has no lid.)

Solution: Back in Chapter 1, "What Is Calculus, Anyway?" I gave you a hint on how to create a box out of a flat piece of paper. Try it for yourself. Place a rectangular sheet of paper in front of you and cut equal squares from the corners. You'll end up with smaller rectangles along the sides of your paper. Fold these up, toward you, along the seam created by the inner sides of the recently removed squares. Can you see how the pieces of rectangle left correspond to dimensions of the box (see Figure 13.5)?

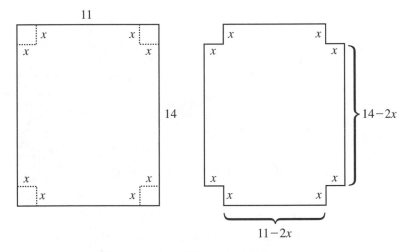

Figure 13.5

The height of the box will be
x inches, since the side of the
square dictates how deeply to
fold the paper.

I have labeled the side of the corner squares as x in Figure 13.5. Once I cut out these squares, the length of the top and bottom is $11 - 2x$, since it was 11 inches and I removed two lengths measuring x inches. Similarly, the sides of the box will measure $14 - 2x$ inches.

Now that you have a good idea what is happening visually, lets get hip-deep in the math. We are trying to make the largest possible volume, so our primary equation should be of the volume for this box. The volume for any box like this is $V = l \cdot w \cdot h$, where l = length, w = width, and h = height. Plug in the correct values for l, w, and h:

$$V = (14 - 2x)(11 - 2x)x$$
$$V = 4x^3 - 50x^2 + 154x$$

Kelley's Cautions

As you plug in for the variables, your goal should be to have only one main variable in the equation. In Example 4, we want to change *l*, *w*, and *h* so they all contain one variable only. Don't worry that *V* is a variable—we won't be dealing with the left side of the equation at all.

If you plug in any *x*, this function gives you the volume of the box, which results from a 11-by-14-inch piece of paper with squares of side *x* cut out. Cool, eh? We want to find what value of *x* makes *V* the largest, so let's find the value guaranteed by the Extreme Value Theorem. Take the derivative with respect to *x* and do a wiggle graph (see Figure 13.6), just like we did in Chapter 11.

$$V' = 12x^2 - 100x + 154 = 0$$
$$6x^2 - 50x + 77 = 0$$

Figure 13.6

The wiggle graph for the first derivative of V. *A relative maximum occurs at* x ≈ 2.039.

Even though *x* = 6.295 appears to be a minimum, the answer doesn't make sense (see the preceding "Kelley's Cautions" sidebar). The maximum volume is reached when *x* = 2.039 (because *V'* changes from positive to negative there, meaning that *V* went from increasing to decreasing), so the optimal dimensions are 2.039 inches by 6.922 inches (11 – 2*x*) by 9.922 inches (14 – 2*x*).

Kelley's Cautions

Without too much figuring, we know that we should only consider values of *x* between 0 and 5.5 for Example 4. Why? Well, if *x* is less than 0, we're not cutting out any squares, and if *x* is greater than 5.5, then the (11 – 2*x*) width of our box becomes 0 or smaller, and that's just not allowed. Our real-life box must have *some* width.

Here are the steps for optimizing functions:

1. Construct an equation in one variable that represents what you are trying to maximize.

2. Find the derivative with respect to the variable in the problem and draw a wiggle graph.

3. Verify your solutions as the correct extrema type (either maximum or minimum) by viewing the sign changes around it in the wiggle graph.

You've Got Problems

Problem 4: What is the minimum product you can achieve from two real numbers, if one of them is three less than twice the other?

The Least You Need to Know

◆ L'Hôpital's Rule is a shortcut to finding limits that are indeterminate when you try to solve them using substitution.

◆ The Mean Value Theorem guarantees that the secant slope on an interval will equal the tangent slope somewhere on this interval—i.e., the average rate of change must somewhere be equal to the instantaneous rate of change.

◆ You can determine how quickly a variable is changing in an equation if you know how quickly the other variables in the equation are changing.

◆ The first derivative can help you determine where a function reaches its optimal values.

Part 4

The Integral

For those of you with a good background in superhero lore, you'll know what I mean by Bizzarro world. In Bizzarro world, everything is the opposite of this world; good means bad, up means down, and right means left. Since Superman is smart in our world, Bizzarro Superman is stupid. Well, integrals are Bizzarro derivatives. Deriving takes us from a function to an expression describing its rate of change, but integrating takes us the opposite direction—from the rate of change back to the original function.

Even though integrating is simply the opposite of deriving, you might think that its usefulness would be limited. You'd be wrong. There are just about as many applications for integrals as there were for derivatives, but they are completely different in nature. Instead of finding rates of change, we'll calculate area, volume, and distance traveled. We'll also explore the Fundamental Theorem of Calculus, which explains the exact relationship between integrals and the area beneath a curve. It's surprising how straightforward that relationship is and how dang useful it can be.

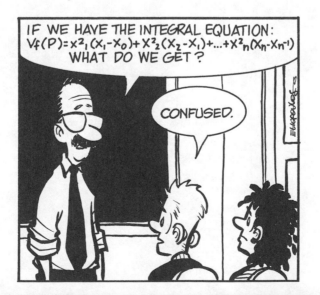

Chapter 14

Approximating Area

In This Chapter

◆ Using rectangles to approximate area

◆ Right, left, and midpoint sums

◆ Trapezoidal approximations

◆ Parabolic approximations with Simpson's Rule

Have you ever seen the movie *Speed* with Keanu Reeves and Sandra Bullock? If not, here's a recap. Everyone's trapped in this city bus, which will explode if the speedometer goes below 50 mph. So, you've got this killer, runaway bus that's flying around the city that can't stop—the perfect breeding ground for destruction, disaster, high drama, mayhem, and a budding romance between the movie's two stars. (Darn that Keanu … talk about being in the right place at the right time ….)

By now, you probably feel like you're on that bus. Calculus is tearing all over the place, never slowing down, never stopping, and (unfortunately) never inhabited by such attractive movie stars. The more you learn about derivatives, the more you have to remember about the things that preceded them. Just when you understand something, another (seemingly unrelated) topic pops up to confound your understanding. When will this dang bus slow down? Actually, the bus slows down now.

You may feel a slight lurching in the pit of your stomach as we slow to a complete stop, and start discussing something completely and utterly different for a while. Until now, we have spent a ton of time talking about rates of change and tangent slopes. That's pretty much over. Instead, we're going to start talking about finding the area under curves. I know—that's a big change, but it'll all be related in the end. For now, take a deep breath, and enjoy a much slower pace for a few chapters as we talk about something different. And if you see Sandra Bullock, tell her I said hi.

Riemann Sums

Let me begin by saying something very deep. Curves are really, really curvy. It is this inherent curviness that makes it hard to find the area beneath them. For example, take a look at the graph of $y = x^2 + 1$ in Figure 14.1 (only the interval [0,3] is pictured).

Figure 14.1

If only the shaded space between $y = x^2 + 1$ and the x-axis were a square or a rectangle—that would make finding the area so much easier.

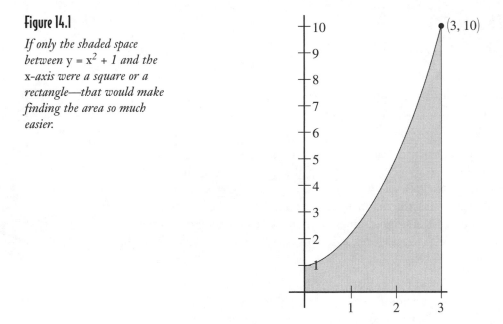

We want to try and figure out exactly how much area is represented by the shaded space. We don't have any formulas from geometry to help us find the area of such a curved figure, so we're going to need to come up with some new techniques. To start with, we're going to approximate that area using figures for which we already have area formulas. Even though it seems kind of lame, we're going to approximate the shaded area using rectangles. The process of using rectangles to approximate area is called *Riemann sums*.

Talk the Talk

A **Riemann sum** is an approximation for the area beneath a curve that is achieved using rectangles. We are using very simple Riemann sums. Some calculus courses will explore very complicated sums, which involve complicated formulas containing sigma signs (Σ). These are a little beyond us, and they really don't help you understand the underlying calculus concepts at all, so we omit them.

Right and Left Sums

I am going to approximate that shaded area beneath $y = x^2 + 1$ using three rectangles. Since I am only finding the area on the x-interval [0,3], that means I will be using three rectangles, each of width 1. (If I had been using six rectangles on an interval of length 3, each rectangle would have width $\frac{1}{2}$.) How high should I make each rectangle? Well, I choose to use a *right sum*, which means that the rectangles will be the height reached by the function at the right side of each interval of length 1, as pictured in Figure 14.2.

Critical Point

When I say we are looking for the area *beneath* the curve, I actually mean the area *between* the curve and the x-axis; otherwise, the area beneath a curve would almost always be infinite. You can always assume that you are finding the area between the curve and the x-axis unless the problem states otherwise.

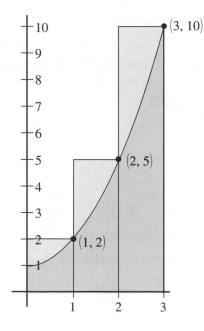

Figure 14.2

I am using three rectangles to approximate the area on [0,3]. The three rectangles cover the intervals [0,1], [1,2], and [2,3].

The rectangle on [0,1] will have the height reached at the far right side of the interval (i.e., $x = 1$), which is 2. Similarly, the second rectangle is 5 units tall, since that is the height of the function at $x = 2$, the right side of its interval. Therefore, the heights of the rectangles are 2, 5, and 10, from left to right. The width of each rectangle is 1.

We can approximate the area beneath the curve by adding the areas of the three rectangles together. Since the area of a triangle is equal to length times width, the total area captured by the rectangles is $1 \cdot 2 + 1 \cdot 5 + 1 \cdot 10 = 17$. Therefore, the right Riemann approximation with $n = 3$ rectangles is 17.

Critical Point _____

It was easy to see that the width of every rectangle in our right sum was 1. If the width of the rectangles is not so obvious, use the width formula $\Delta x = \frac{b-a}{n}$ to calculate the width. In this formula, the interval $[a,b]$ is being split up into n different rectangles, and each will have width Δx. In our right sum example, $\Delta x = \frac{3-0}{3} = 1$, since we are splitting up the interval $[0,3]$ into $n = 3$ rectangles.

Talk the Talk _____

The kind of sum you're calculating depends on how high you make the rectangles. If you use the height at each rectangle's left boundary, you're finding **left sums.** If you use the height at the right boundary of each rectangle, the result is **right sums.** Obviously, **midpoint sums** use the height reached by the function in the middle of each interval.

Clearly, the area captured by the rectangles is much more than is beneath the curve. In fact, it looks like a lot more. This should tell you that we have got to come up with better methods later (and we indeed will). For now, let's have a go at the same area problem, but this time use four rectangles and *left sums.*

Example 1: Approximate the area beneath the curve $f(x) = x^2 + 1$ on the interval [0,3] using a left Riemann sum with four rectangles.

Solution: To find how wide each of the four rectangles will be, use the formula $\Delta x = \frac{b-a}{n}$:

$$\Delta x = \frac{3-0}{4} = \frac{3}{4}$$

If each of the four intervals is $\frac{3}{4}$ wide, and the rectangles start at 0, then the rectangles will be defined by the intervals $\left[0,\frac{3}{4}\right], \left[\frac{3}{4},\frac{3}{2}\right], \left[\frac{3}{2},\frac{9}{4}\right],$ and $\left[\frac{9}{4},3\right]$. (This is because $0 + \frac{3}{4} = \frac{3}{4},$ $\frac{3}{4} + \frac{3}{4} = \frac{6}{4} = \frac{3}{2}$, etc.) You will be using the heights reached by the function at the left boundary of each interval. Therefore, the heights will be $f(0), f\left(\frac{3}{4}\right), f\left(\frac{3}{2}\right),$ and $f\left(\frac{9}{4}\right)$ in Figure 14.3.

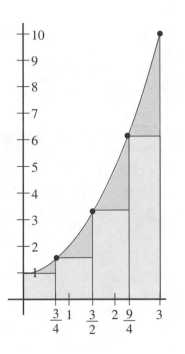

Figure 14.3

Each of the four rectangles is $\frac{3}{4}$ wide, and they are as high as the function $f(x) = x^2 + 1$ is at the left edge of each rectangle, hence left sums is the result.

The area of each rectangle is its width times its height, so the total area is …

$$\frac{3}{4} \cdot f(0) + \frac{3}{4} \cdot f\left(\frac{3}{4}\right) + \frac{3}{4} \cdot f\left(\frac{3}{2}\right) + \frac{3}{4} \cdot f\left(\frac{9}{4}\right)$$

$$\frac{3}{4} \cdot 1 + \frac{3}{4} \cdot \frac{25}{16} + \frac{3}{4} \cdot \frac{13}{4} + \frac{3}{4} \cdot \frac{97}{16} = \frac{285}{32} = 8.90625$$

This number underestimates the actual area under the curve, since there are large pieces of area under the curve missed by our rectangles.

Midpoint Sums

Calculating midpoint sums is very similar to right and left sums. The only difference is (you guessed it) how you define the heights of the rectangles. In our ongoing example of $f(x) = x^2 + 1$ on the x-interval [0,3], let's say we wanted to calculate midpoint sums using (to make it easy) $n = 3$ rectangles. As before, the intervals defining the rectangles' boundaries will be [0,1], [1,2], and [2,3], and each rectangle will have a width of 1. What about the heights?

Look at the interval [0,1]. If we were using left sums, the height of the rectangle would be $f(0)$. If using right sums, it'd be $f(1)$. However, we're using midpoint sums, so you use the function value at the *midpoint* of the interval, which in this case is $\frac{1}{2}$. Therefore, the height of the rectangle is $f\left(\frac{1}{2}\right)$. If you apply this to all three intervals, the midpoint Riemann approximation of the area would be …

$$1 \cdot f\left(\frac{1}{2}\right) + 1 \cdot f\left(\frac{3}{2}\right) + 1 \cdot f\left(\frac{5}{2}\right) = 11.75$$

You've Got Problems

Problem 1: Approximate the area beneath the curve $g(x) = -\cos x$ on the interval $\left[\frac{\pi}{2}, \frac{3\pi}{2}\right]$ using $n = 4$ rectangles and (1) left sums, (2) right sums, and (3) midpoint sums.

The Trapezoidal Rule

All of our techniques so far are pretty rough when it comes to approximating. The Trapezoidal Rule, however, is a much more accurate way to approximate area beneath a curve. Instead of constructing rectangles, this method uses small trapezoids. In effect, these trapezoids look the same as their predecessor rectangles near their bases, but completely different at the top. To construct the trapezoids, you mark the height of the function at the beginning and end of the width interval (which is still calculated by the formula $\Delta x = \frac{b-a}{n}$) and connect those two points. Figure 14.4 is a diagram representing how the Trapezoidal Rule approximates the area beneath our favorite function in the whole world, $y = x^2 + 1$.

Critical Point

If you're dying to know the actual area beneath $y = x^2 + 1$ on the interval $[0,3]$, it is exactly 12. Of our techniques so far, the midpoint sum came the closest (even though we used only three rectangles with this method but four with left sums).

There's a lot less room for error with this rule, and it's actually just as easy to use as Riemann sums were. One difference—this one requires that you memorize a formula.

Critical Point

This is going to freak you out. Remember how the left and right sums offset one another when we approximated the area beneath $y = x^2 + 1$—one too big and the other way too small? Well, the Trapezoidal Rule (with n trapezoids) is exactly the average of the left and right sums (with n rectangles). We already know that the right sum of $y = x^2 + 1$ (with $n = 3$) is 17. You can find the matching left sum to be 7. If you calculate the Trapezoidal Rule approximation (with $n = 3$ trapezoids), you get 12, which is the average of 7 and 17.

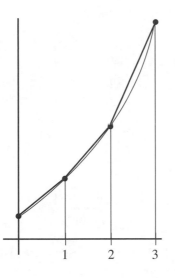

Figure 14.4

The "top" of our shape is no longer parallel to the x-axis. Instead, it connects the function's heights at the interval endpoints.

The Trapezoidal Rule: The approximate area beneath a curve $f(x)$ on the interval $[a,b]$ using n trapezoids is …

$$\frac{b-a}{2n}\left(f(a)+2f(x_1)+2f(x_2)+2f(x_3)+\ldots+2f(x_{n-1})+f(b)\right)$$

In practice, you pop the correct numbers into the fraction at the beginning and then evaluate the function at every interval boundary. Except for the endpoints, you'll multiply all the values by 2. Let's go straight into an example, and you'll see that it's not very hard at all. Just for grins, let's use the same function we have been and see if the Trapezoidal Rule can beat out our current best estimate of 11.75 given by the midpoint sum.

The area of any trapezoid is one-half of the height times the sum of the bases (the bases are the parallel sides). For the trapezoid in Figure 14.5, the area is $\frac{1}{2}h(b_1+b_2)$. You may not be used to seeing trapezoids tipped on their side like this—in geometry, the bases are usually horizontal, not vertical. The reason you see all those 2's in the Trapezoidal Rule is that every base is used twice for consecutive trapezoids except for the bases at the endpoints.

Critical Point

There is another way to get better approximations using Riemann sums. If you increase the number of rectangles you use, the amount of error decreases. However, the amount of calculating you have to do increases. Eventually, we'll find a way to obtain the *exact* area without much work at all. It's actually rooted in Riemann sums, but uses an *infinite* number of rectangles in order to eliminate any error completely.

Figure 14.5

Our approximation trapezoids are simply right trapezoids shoved onto their sides, with bases b_1 and b_2 and height h.

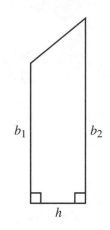

Kelley's Cautions

Even though the Trapezoidal Rule's formula contains the expression $\frac{b-a}{2n}$, you still use the formula $\frac{b-a}{n}$ to find the width of the trapezoids. Don't get them confused—they are separate formulas.

Example 2: Approximate the area beneath $f(x) = x^2 + 1$ on the interval [0,3] using the Trapezoidal Rule with $n = 5$ trapezoids.

Solution: Since we are using five trapezoids, we need to determine how wide each will be, so use the Δx formula:

$$\Delta x = \frac{b-a}{n} = \frac{3-0}{5} = \frac{3}{5}$$

Therefore, the boundaries of the intervals will start at $x = 0$ and progress in steps of $\frac{3}{5}$: 0, $\frac{3}{5}$, $\frac{6}{5}$, $\frac{9}{5}$, $\frac{12}{5}$, and 3. These numbers belong in the formula as a, x_1, x_2, x_3, x_4, and b. So, according to the Trapezoidal Rule, the area is approximately:

$$\frac{3-0}{2(5)}\left(f(0) + 2f\left(\frac{3}{5}\right) + 2f\left(\frac{6}{5}\right) + 2f\left(\frac{9}{5}\right) + 2f\left(\frac{12}{5}\right) + f(3)\right)$$

$$= \frac{3}{10}\left(1 + 2\cdot\frac{34}{25} + 2\cdot\frac{61}{25} + 2\cdot\frac{106}{25} + 2\cdot\frac{169}{25} + 10\right)$$

$$= \frac{609}{50} \approx 12.18$$

This is actually our closest approximation yet, although it is a bit over. Had this curve been concave down instead of up, the result would have underestimated the area. Can you see why? The teeny bit of error would have been outside, rather than inside, the curve.

> ### You've Got Problems
>
> Problem 2: Approximate the area beneath $y = \sin x$ on the interval $[0,\pi]$ using the Trapezoidal Rule with $n = 4$ trapezoids.

Simpson's Rule

Our final area-approximating tool is Simpson's Rule. Geometrically, it creates tiny little parabolas (rather than the slanted but still linear tops of the trapezoids) to wrap even closer around the function we're approximating. The formula is astonishingly similar to the Trapezoidal Rule, but here's the catch: You can only use an even number of subintervals.

Simpson's Rule: The approximate area under the curve $f(x)$ on the closed interval $[a,b]$ using an even number of subintervals, n, is …

$$\frac{b-a}{3n}\left(f(a) + 4f(x_1) + 2f(x_2) + \ldots + 2f(x_{n-2}) + 4f(x_{n-1}) + f(b)\right)$$

In this formula, the outermost terms get multiplied by nothing. However, beginning with the second term, you multiply consecutive terms by 4, then 2, then 4, then 2, etc. Make sure you always start with 4, though. Back to Old Faithful for an example one more time.

Example 3: Approximate the area beneath that confounded function $y = x^2 + 1$ on the closed interval $[0,3]$ one more time, this time using Simpson's Rule and $n = 6$ subintervals.

Solution: Some quick calculating tells us that our subintervals will have the width of $\Delta x = \frac{b-a}{n} = \frac{3-0}{6} = \frac{1}{2}$. Now, into the formula we go:

$$\frac{b-a}{3n}\left(f(0) + 4f\left(\frac{1}{2}\right) + 2f(1) + 4f\left(\frac{3}{2}\right) + 2f(2) + 4f\left(\frac{5}{2}\right) + f(3)\right)$$

Remember to multiply $f(\frac{1}{2})$ by 4, the next term by 2, etc. However, the last term gets no additional coefficient:

$$\frac{3-0}{3\cdot 6}\left(1 + 4\cdot\frac{5}{4} + 2\cdot 2 + 4\cdot\frac{13}{4} + 2\cdot 5 + 4\cdot\frac{29}{4} + 10\right)$$

$$= \frac{1}{6}\left(1 + 5 + 4 + 13 + 10 + 29 + 10\right) = 12$$

Whoa! Since Simpson's Rule uses quadratic approximations, and this is a quadratic formula, we get the exact answer. This only happens for areas beneath quadratic equations, though.

Problem 3: Approximate the area beneath $y = \frac{1}{x}$ on the interval $[1,5]$ using Simpson's Rule with $n = 4$ subintervals.

The Least You Need to Know

◆ Riemann sums use rectangles to approximate the area beneath a curve; the heights of these rectangles are based on the height of the function at the left end, right end, or midpoint of each subinterval.

◆ The width of each subinterval in all the approximating techniques is $\Delta x = \frac{b-a}{n}$.

◆ The Trapezoidal Rule is the average of the left and right sums, and usually gives a better approximation than either does individually.

◆ Simpson's Rule uses intervals topped with parabolas to approximate area; therefore, it gives the exact area beneath quadratic functions.

Antiderivatives

In This Chapter

- ◆ "Un-deriving" expressions
- ◆ The Power Rule for Integration
- ◆ Integrating trigonometric functions
- ◆ The Fundamental Theorem: the connection to area
- ◆ The key to *u*-substitution

Are you a little perplexed? Probably. We spent the first 13 chapters of the book discussing complex mathematical procedures, and then suddenly and without warning, we're calculating the area of rectangles in Chapter 14. Kind of a letdown, I know. Most people have this terrifying view of calculus, and assume that everything in it is impossible to understand; they are usually surprised to be calculating simple areas this deep in the course.

In this chapter, we'll find *exact* areas beneath a curve. We'll also uncover one of the most fascinating mathematical relationships of all time: The area beneath a curve is related to the curve's antiderivative. You heard me right—antiderivative. After all this time learning how to find the derivative of a function, now we're going to go backward and find the antiderivative. Before, we took $f(x) = x^3 - 2x^2$ and got $f'(x) = 3x^2 - 4x$; now, we're going to start with the derivative and figure out the original function.

It's a whole new ballgame, and we're going to learn everything from the first half of the course in reverse. For those of us who always seem to do things backwards, this should come as a welcome change! Sound exciting? Sound painful? It's a little from column A and a little from column B.

The Power Rule for Integration

Before we get started, let's talk briefly about what reverse differentiating means. The process of going from the expression $f'(x)$ back to $f(x)$ is called *antidifferentiation* or *integration*—both words mean the same thing. The result of the process is called an *antiderivative* or an *integral*. Integration is denoted using a long, stretched out letter S, like this:

$$\int 2x\,dx = x^2 + C$$

This is read "The integral of $2x$, with respect to x, is equal to x^2 plus some unknown constant (called the *constant of integration*)." This integral expression is called an *indefinite integral* since there are no boundaries on it, whereas an expression such as $\int_1^3 2x\,dx$ is called a *definite integral*, because it contains the *limits of integration* 1 and 3. Both expressions contain a "*dx*"; don't worry at all about this little piece—you don't have to do anything with it. Just make sure its variable matches the variable in the function (in this case, x).

Talk the Talk

The opposite of a derivative is called an **antiderivative** or **integral**. If $f(x)$ is the antiderivative of $g(x)$, then $\int g(x)dx = f(x)$. The process of creating such an expression is called **antidifferentiation** or **integration**. Why do you have to use a **constant of integration?** Lots of functions have the same derivative—for example, both $f(x) = x^3 + 6$ and $f(x) = x^3 - 12$ have the same derivative, $3x^2$. Therefore, when we integrate $\int 3x^2 dx$, we say the antiderivative is $x^3 + C$, since we have no way of knowing what constant was in the original function. All indefinite integrals must contain a "$+C$" in their solution for this reason. An **indefinite integral** has no boundaries next to the integration sign. Its solution is an antiderivative. A **definite integral** has boundaries, called **limits of integration,** next to the integration sign. Its solution is a real number. For example, $\int 2x\,dx = x^2 + C$, but $\int_1^3 2x\,dx = 8$. (You'll learn how to do both of these procedures soon.)

That's a lot of vocabulary for now. Before you get overwhelmed, let's get into the meat of the mathematics. Remember finding simple derivatives with the Power Rule? There's a way to find simple integrals using the *Power Rule for Integration*. Instead of multiplying the original coefficient by the exponent and then subtracting 1 from the power, you'll *add* 1 to the power and *divide* by the new power.

The Power Rule for Integration: The integral of a single variable to some power is found by adding 1 to the existing exponent and dividing the entire expression by the new exponent:

$$\int x^n \, dx = \frac{x^{n+1}}{n+1} + C$$

Remember: You can only use the Power Rule for Integration if integrating a single variable to a power, just like the regular Power Rule. However, if the only thing standing in your way is a coefficient, you are allowed to pull that out of the integral to get it out of your way, as indicated in the first example.

Example 1: Evaluate $\int \left(7x^3 + 6x^5\right) dx$.

Solution: Even though there are two terms here, each is simply a variable to some power with a coefficient attached. You can actually separate addition or subtraction problems into separate integrals as follows:

$$\int 7x^3 \, dx + \int 6x^5 \, dx$$

Don't worry about the $\int$ or dx in the problem. You just want to integrate whatever falls between them. Before you can apply the Power Rule for Integration, you should "pull out" the coefficients:

$$7\int x^3 \, dx + 6\int x^5 \, dx$$

Now the expression in the integral looks like the one in the Power Rule for Integration theorem. Add 1 to each power and divide the new expressions by that power. The integral sign and the "dx" will disappear, but don't forget to add a "$+C$" to the end of the problem, since all indefinite integrals require it:

$$7 \cdot \frac{x^4}{4} + 6 \cdot \frac{x^6}{6} + C$$
$$\frac{7}{4}x^4 + x^6 + C$$

Critical Point

You pull the coefficients out of the integrals to make the integration itself easier. As soon as the integration sign is gone, you end up multiplying that coefficient by the integral anyway, so it's not as though it "goes away" somewhere. It just hangs around, waiting for the integration to be done.

Integrating Trigonometric Functions

Just like learning trigonometric derivatives, learning trigonometric integrals is just a question of memorizing the correct quantity. These should become second nature to you, and you should be able to recite them without a second thought. If you forget them, you can actually create some of them from scratch easily (like the integral of the tangent function, as you'll see later in the chapter). However, not all are quite so easy to recreate, so I see some quality memorizing time in your not-too-distant future.

Critical Point

All of the integral solutions on the list containing a "co-" function are negative.

I can tell by that pouty look on your face that the thought of more memorizing doesn't excite you. (You're going to get even more pouty if you haven't flipped ahead to the actual formulas yet—they are crazy-looking.) Think back. You had to memorize the multiplication tables in elementary school, remember? This is just sort of the grandfather of the multiplication tables, but important all the same.

And now, with no further ado here are the trigonometric functions with their anti-derivatives:

- $\int \sin x\, dx = -\cos x + C$
- $\int \cos x\, dx = \sin x + C$
- $\int \tan x\, dx = -\ln|\cos x| + C$
- $\int \cot x\, dx = \ln|\sin x| + C$
- $\int \sec x\, dx = \ln|\sec x + \tan x| + C$
- $\int \csc x\, dx = -\ln|\csc x + \cot x| + C$

There are a lot of natural log functions in the list of trig integrals. That is due, in no small part, to the fact that $\int \frac{1}{x}\, dx = \ln|x| + C$, another important formula to memorize.

Here's another, while we're at it: The integral of e^x is itself, just like it was its own derivative; therefore, $\int e^x dx = e^x + C$. Integrating logarithmic functions is very, very tricky, so we don't even attempt that until Chapter 18. Hey, now you have something to look forward to!

The Fundamental Theorem of Calculus

Finally, it's time to solve two mysteries of recent origin: (1) How do you find the exact areas under curves, and (2) Why are we even mentioning areas—isn't this chapter about integrals? It turns out that the exact area beneath a curve can be computed using a definite integral. This is one of two major conclusions, which together make up the Fundamental Theorem.

Part One: Areas and Integrals Are Related

After all this time, we're going to finally be able to show that the area beneath the curve $y = x^2 + 1$ on the interval [0,3] is equal to 12. We spent most of Chapter 14 trying to estimate the area, and now we're finally going to be able to do it. Huzzah!

From now on, we're going to equate definite integrals with the area beneath a curve (technically speaking, the area between the function and the x-axis, remember?). Therefore, I can say that the area beneath $x^2 + 1$ on the interval [0,3] is equal to $\int_0^3 \left(x^2 + 1 \right) dx$.

This new notation is read, "the integral of $x^2 + 1$, with respect to x, from 0 to 3." Unlike indefinite integrals, the solution to a definite integral, such as this one, is a number. That number is, in fact, the area beneath the curve. How in the world do you get that number, you ask? How about a warm welcome for the Fundamental Theorem?

The Fundamental Theorem of Calculus (part one):
If $g(x)$ is any antiderivative of the continuous function $f(x)$, then $\int_a^b f(x)dx = g(b) - g(a)$.

In other words, to calculate the area beneath the curve $f(x)$ on the interval [a,b], you must first integrate the function. Then, plug the upper bound (b) into the integral. From this value, subtract the result you get from plugging the lower bound (a) into the same integral. It's a brilliantly simple process, as powerful as it is elegant.

Critical Point

You will get a negative answer from a definite integral if the area in question is below the x-axis. Whereas "negative area" does not make sense, we must automatically assign all area below the x-axis with a negative value.

Critical Point _____

Here are two important properties of definite integrals to be aware of:

♦ $\int_a^a f(x)dx = 0$ (i.e., if the upper and lower limits of integration are equal, the definite integral equals 0)

♦ $\int_a^b f(x)dx = -\int_b^a f(x)dx$ (i.e., you can switch the limits of integration if you like—just pop a negative sign out front and everything's kosher)

Example 2: Once and for all, find the *exact* area beneath the curve $f(x) = x^2 + 1$ on the interval $[0,3]$ using the Fundamental Theorem of Calculus.

Solution: This problem asks us to evaluate the definite integral.

$$\int_0^3 \left(x^2 + 1\right)dx$$

Begin by integrating $x^2 + 1$ using the power rule for integration. When you complete the integral, you no longer write the integration symbol, and you do not write "$+C$." Instead, draw a vertical slash to the right of the integral, and copy the limits of integration onto it. This signifies that the integration portion of the problem is done:

$$\left(\frac{x^3}{3} + x\right)\Big|_0^3$$

Now, plug 3 into the function (for both x's) and subtract 0 plugged into the function:

$$\left(\frac{3^3}{3} + 3\right) - \left(\frac{0^3}{3} + 0\right) = 9 + 3 = 12$$

You've Got Problems

Problem 2: Calculate $\int_{\pi/2}^{3\pi/2} \cos x \, dx$. Explain what is meant by the answer.

Part Two: Derivatives and Integrals Are Opposites

I kind of spoiled this revelation for you already; I'm sorry. However, the second major conclusion of the Fundamental Theorem still holds some surprises. Let's look at the mathematical wording of the theorem first:

The Fundamental Theorem of Calculus (part two): If $f(x)$ is a continuous and differentiable function, $\frac{d}{dx}\left(\int_a^{f(x)} g(y)dy\right) = g(f(x)) \cdot f'(x)$, if a is a real number.

That looks unsightly. Here's what it means without all the gobbledygook. Let's say you're taking the derivative of a definite integral whose lower bound is a constant (i.e., just a number) and whose upper bound contains a variable. If you take the derivative of the integral expression with respect to the variable present in the upper bound, the answer will be the function inside the integral sign (unintegrated), with the upper bound plugged in, multiplied by the derivative of the upper bound. This theorem looks, feels, and even smells complex, but it's not hard at all. Trust me on this one. All you have to do is learn the pattern.

Example 3: Evaluate $\frac{d}{dx}\left(\int_{\sin x}^{3} t^2 dt\right)$.

Solution: You don't *have* to use the shortcut in part two of the Fundamental Theorem, but it makes it easier. Notice that the variable expression is in the lower (not the upper) bound, which is not allowed by the theorem. Therefore, I will swap them using a property of integrals we've already discussed (which makes the integral negative):

$$\frac{d}{dx}\left(-\int_3^{\sin x} t^2 dt\right)$$

Since we're deriving with respect to x (and x is in the upper bound) and the lower bound is a constant, we're clear to apply our new theorem. All you do is plug the upper bound (sin x) into the function t^2 to get (sin $x)^2$, and multiply by the derivative of the upper bound (which will be cos x). Don't forget the negative, which stays out in front of everything:

$$-\sin^2 x \cdot \cos x$$

Critical Point

What if you forget this theorem? No problem—you can do Example 3 another way, working from the inside out (i.e., start with the integration problem and then take the derivative). You get the same thing. If you apply the Fundamental Theorem part one to the integral, you get:

$$\frac{d}{dx}\left(\frac{t^3}{3}\Big|_{\sin x}^{3}\right) = \frac{d}{dx}\left(9 - \frac{1}{3}\sin^3 x\right)$$

Take the derivative with respect to x to get $-\sin^2 x \cdot \cos x$. Don't forget to apply the Chain Rule when deriving $\frac{1}{3}\sin^3 x$—that's how you get the $\cos x$.

You don't always have to switch the boundaries and make the integral negative. This must only be done if the constant appears in the wrong boundary. What happens if both boundaries contain variables? If this is the case, you cannot use the shortcut offered by the theorem and must resort to the long way described in the preceding sidebar.

You've Got Problems

Problem 3: Evaluate $\frac{d}{dx}\left(\int_1^{\tan x} e^t \, dt\right)$ twice, once using the Fundamental Theorem of Calculus part one, and once using part two.

U-Substitution

The only integration techniques you know so far are the Power Rule and the functions whose integrals you memorized (like cos x and e^x). The key to u-substitution, our next integration technique, is finding a piece of the function whose derivative is also in the function. The derivative is allowed to be off by a coefficient, but otherwise must appear in the function itself.

Here are the steps you'll follow when u-substituting. First, look for a piece of the function, as I've mentioned, whose derivative is also in the function. If you're not sure what to use, try the denominator or something being raised to a power in the function. Second, set u equal to that piece of the function and take the derivative with respect to nothing. Lastly, use your u and du expressions to replace parts of the original integral, and your new integral will be much easier to solve.

Example 4: Use u-substitution to find $\int \frac{\sin x}{\cos x} \, dx$ (i.e., prove that the integral of tangent is

$-\ln|\cos x| + C$).

Critical Point

Deriving with respect to nothing means following the derivative with a "dx," "dt," or similar term, depending on the variable in the expression. For example, the derivative of sin$x$ with respect to nothing is cos x dx (just add a "dx"). The derivative of ln t with respect to nothing is $\frac{1}{t} \, dt$. You derive the same but just add the extra piece at the end.

Solution: As the problem hints, we already know the integral of $\frac{\sin x}{\cos x}$ (which, by the way, equals tan$x$). We're just going to prove it. I have to set u equal to a piece of the problem whose derivative is also there. Well, since sine and cosine are both present (and the derivative of each is basically the other function), I could pick either one to be u, but remember the hint I gave you: If you're not sure, choose the denominator or something to a power. Therefore, I set $u = \cos x$ and derive with respect to nothing to get $du = -\sin x \, dx$. Notice that sin$x$ dx is in the problem, so solve for sin$x$ dx in the du equation by dividing by -1; you get $-du = \sin x \, dx$. Now, go back to the original integral and replace cos$x$ with u (since we set $u = \cos x$) and replace sin$x$ dx with $-du$ (since $-du = \sin x \, dx$):

$$\int \frac{-du}{u} = -\int \frac{du}{u}$$

Talk the Talk

You must use *u*-substitution to integrate a function containing something other than just x, just as you had to use the Chain Rule to differentiate such functions. For example, in the integral $\int \cos(3x)dx$, the cosine function contains $3x$, not just x, so set $u = 3x$ and $du = 3 \cdot dx$. Only dx is in the function (not $3dx$), so solve for it to get $\frac{du}{3} = dx$. These changes make the integral $\int \cos u \cdot \frac{du}{3}$, or $\frac{1}{3}\int \cos u \, du$, which integrates as $\frac{1}{3}\sin 3x + C$.

Remember that the integral of $\frac{1}{x}$ is $\ln|x|$, so this integral becomes $-\ln|u| + C$. Finally, replace the u using your original u equation ($u = \cos x$) to get the final answer of $-\ln|\cos x| + C$.

The trickiest part of u-substitution is deciding what u should be. If your first choice doesn't work, don't sweat it. Try something else until it works out for you. It eventually will. The only way to get really good at this method is through practice, practice, practice. Eventually, picking the correct u will become easier.

You've Got Problems

Problem 4: Evaluate $\int_0^{\pi/4} \sec^2 x \tan x \, dx$. Hint: If you are performing u-substitution with a definite integral, you have to change the limits of integration as you substitute in the u and du statements. To change the limits, plug them each into the x slot of your u equation and replace the original x limits with each result.

The Least You Need to Know

◆ Integration, like differentiation, has a Power Rule all its own, in which you add 1 to the exponent and divide by the new exponent.

◆ Trigonometric functions have bizarre integrals, some of which are difficult to produce on your own; therefore, it's best to memorize them.

◆ The two parts of the Fundamental Theorem of Calculus tell you how to evaluate a definite integral and give a shortcut for finding specific derivatives of integral expressions.

◆ U-substitution helps you integrate expressions that contain functions and their derivatives.

Applications of the Fundamental Theorem

In This Chapter

◆ Finding yet more curvy area

◆ Integration's Mean Value Theorem

◆ Position equations and distance traveled

◆ Functions defined by definite integrals

Once we learned how to find the slope of a tangent line (a seemingly meaningless skill), it seemed as though the applications for the derivative would never stop. We were finding velocity and rates of change (both instantaneous and average), calculating related variable rates, optimizing functions, determining extrema, and, all in all, bringing peace and prosperity to the universe.

If you think that the applications for definite integrals are going to start pouring in, you must be psychic. (Either that or you read the table of contents.) For now, we'll look at some of the most popular definite integral-related calculus topics. We'll start by finding area bounded by two curves (rather than one curve and the x-axis). Then, we'll briefly backtrack to topics we've already discussed, but we'll spice them up a little with what we now know of integrals. Finally, we'll look at definite integral functions, also called accumulation functions.

Calculating Area Between Two Curves

This'll blow your mind. In fact, when you read it, your very sanity may be called into question. The thin ribbons of consciousness tying you to this mortal world may stretch and break, catapulting you into madness, or at least making you lose your appetite. Perhaps you should sit down before you continue.

You've been calculating the area between curves *all along without even knowing it*. There, I said it. I hope you're okay.

> ### Critical Point
>
> If you have functions containing *y* instead of functions containing *x* (i.e., $f(y) = y^2$), you can still calculate the area between the functions. However, instead of subtracting top-bottom inside the integral, you subtract right-left.

If you want to calculate the area between two continuous curves, we'll call them $f(x)$ and $g(x)$, on the same *x*-interval $[a,b]$, here's what you do. Set up a definite integral as we did last chapter, with *a* and *b* as the lower and upper limits of integration respectively. Inside the integral will be either $f(x) - g(x)$ or $g(x) - f(x)$. To decide which one to use, you have to graph the functions—you should subtract the lower graph from the higher graph. For example, in Figure 16.1, *g* is below *f* on the interval $[a,b]$.

Figure 16.1

At least on the interval [a,b], *the graph of* f(x) *is always higher than the graph of* g(x).

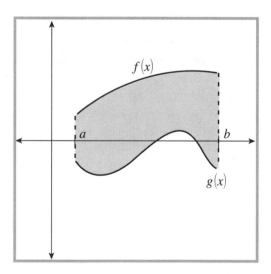

The only place to make a mistake in our new method is subtracting in the wrong order. Doing so will give you a negative answer, and you should *never* get a negative answer when finding the area between curves, even if some of that area falls below the *x*-axis.

What if the curves switch places? For example, look at the graph in Figure 16.2. The curves switch places at $x = c$.

To find the shaded area, we'll have to use two separate definite integrals, one for the interval $[a,c]$, when f is on top, and one for $[c,b]$, when g is: $\int_a^c \big(f(x) - g(x)\big)dx + \int_c^b \big(g(x) - f(x)\big)dx$.

Example 1: Calculate the area between the functions $f(x) = \sin 2x$ and $g(x) = \cos x$ on the interval $\left[\frac{3\pi}{2}, 2\pi\right]$.

Solution: These graphs play leapfrog all along the x-axis, but on the interval $\left[\frac{3\pi}{2}, 2\pi\right]$, $g(x)$ is definitely above $f(x)$ (see Figure 16.3).

> **Critical Point** _____
>
> The reason we've technically been doing this all along is that we've always been finding the area between the curve and the x-axis, which has the equation $g(x) = 0$. Thus, if a function $f(x)$ is above the x-axis on $[a,b]$, the area beneath the two curves is $\int_0^3 \big((x^2 + 1) - 0\big)dx$. That second equation with a value of 0 has been invisible all this time.

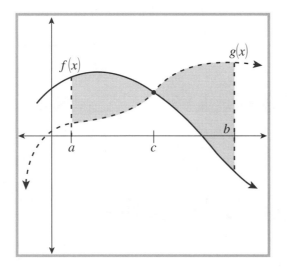

Figure 16.2

The graphs take turns on the top bunk—neither is above the other on the entire interval.

Therefore, our integral will contain $g(x) - f(x)$: $\int_{3\pi/2}^{2\pi} (\cos x - \sin 2x)dx$. Split this up into separate integrals. The first is easy:

$$\int_{3\pi/2}^{2\pi} \cos x \, dx = \big(\sin x\big)\Big|_{3\pi/2}^{2\pi} = \sin 2\pi - \sin \frac{3\pi}{2} = 0 - (-1) = 1$$

You have to use u-substitution to integrate $\sin 2x$, with $u = 2x$:

$$\frac{1}{2}\int_{3\pi}^{4\pi} \sin u \, du = \frac{1}{2}\big(-\cos u\big)\Big|_{3\pi}^{4\pi} = \frac{1}{2}\big(-\cos 4\pi - (-\cos 3\pi)\big) = \frac{1}{2}(-2) = -1$$

Figure 16.3

On the interval $\left[\frac{3\pi}{2}, 2\pi\right]$,

g(x) = cos x *rises above*
f(x) = sin 2x.

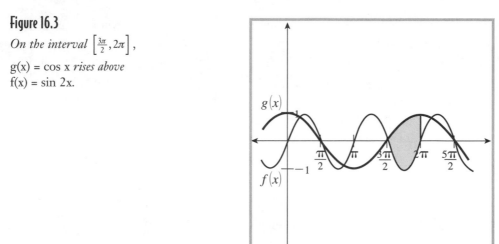

Don't forget to change your x boundaries into u boundaries when you u-substitute. For example, to get the new u boundary of 4π, I plugged the old x boundary of 2π into the u equation: $u = 2(2\pi) = 4\pi$. Our final answer is the first integral minus the second: $1 - (-1) = 2$.

You've Got Problems

Problem 1: Calculate the area between the curves $y = x^2$ and $y = x^3$ in the first quadrant.

The Mean Value Theorem for Integration

Think back to the original Mean Value Theorem for a second. It said that somewhere on an interval, the derivative was equal to the average rate of change for the whole interval. It turns out that integration has its own version of a Mean Value Theorem, but since integration involves area instead of rates of change, it's a bit different.

A Geometric Interpretation

In essence, the Mean Value Theorem for Integration states that at some point along an interval $[a,b]$, there exists a certain point $(c, f(c))$ between a and b (see Figure 16.4). If you draw a rectangle whose base is the interval $[a,b]$ and whose height is $f(c)$, the area of that rectangle will be exactly the area beneath the function on $[a,b]$.

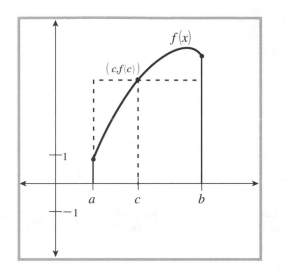

Figure 16.4

A visual representation of the Mean Value Theorem for Integration. The area of the rectangle, whose height is f(c), is exactly equal to

$$\int_a^b f(x)dx.$$

The Mean Value Theorem for Integration: If a function $f(x)$ is continuous on the interval $[a,b]$, then there exists a c, $a \le c \le b$, such that $(b-a) \cdot f(c) = \int_a^b f(x)dx$.

This Mean Value Theorem, like its predecessor, is only an existence theorem. It guarantees that the value $x = c$ and the corresponding key height $f(c)$ exist. You may wonder why it's so important that a curvy graph and a plain old rectangle must always share the same area. We'll get to that after the next example.

Example 2: Find the value $f(c)$ guaranteed by the Mean Value Theorem for Integration for the function $f(x) = x^3 - 4x^2 + 3x + 4$ on the interval $[1,4]$.

Solution: The Mean Value Theorem for Integration states that there is a c between a and b so that $(b-a) \cdot f(c) = \int_a^b f(x)dx$. We know everything except what c is, but that's okay. The problem only asks us to find $f(c)$. Plug in everything you know:

$$(4-1) \cdot f(c) = \int_1^4 \left(x^3 - 4x^2 + 3x + 4\right)dx$$

After the quick subtraction problem on the left (and the slightly lengthier definite integral on the right), you should get ...

Critical Point

In the Mean Value Theorem for Integration statement $(b-a) \cdot f(c) = \int_a^b f(x)dx$, $b - a$ represents the length of the rectangle, since it is the length of the interval $[a,b]$. The height of the rectangle is, as we've already discussed, $f(c)$. There may be more than one such c in the interval, but there must be at least one.

$$3f(c) = \frac{57}{4}$$

$$f(c) = \frac{57}{12} = 4.75$$

This means that the area beneath the curve $f(x) = x^3 - 4x^2 + 3x + 4$ on the interval [1,4] (which equals 14.25, by the way) is equal to the area of the rectangle whose length is the same as the interval's length (3) and whose height is 4.75.

You've Got Problems

Problem 2: Find the value *f(c)* guaranteed by the Mean Value Theorem for Integration on the function $f(x) = \frac{\ln x}{x}$ on the interval [1,100].

The Average Value Theorem

The value *f(c)* that you found in both Example 2 and Problem 2 has a special name. It is called the *average value* of *f*. If we take the equation for the Mean Value Theorem for Integration and divide it by (*b* – *a*), we solve the equation for the average value:

$$f(c) = \frac{\int_a^b f(x)dx}{b-a}$$

This is simply a different form of our previous equation, so it doesn't warrant a whole lot more attention. However, some textbooks completely skip over the Mean Value Theorem for Integration and go right to this, which they call the Average Value Theorem. They might word Problem 2 above as follows: "Find the *average value* of $f(x) = \frac{\ln x}{x}$ on the interval [1,100]." You'd solve the problem the exact same way (see Figure 16.5).

Talk the Talk

The **average value** of a function is the value *f(c)* guaranteed by the Mean Value Theorem for Integration (the height of the rectangle of equal area). The average value is found via the equation $f(c) = \frac{\int_a^b f(x)dx}{b-a}$. Think of it this way. Most functions twist and turn throughout their domains. If you could "level out" a function by filling in its valleys and flattening out its peaks until the function was a horizontal line, the height of that line (i.e., its *y*-value) would be the average value for that function.

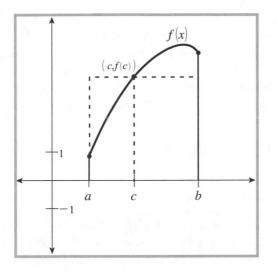

Figure 16.5

Here's the diagram for the Mean Value Theorem for Integration once more. The height of the dotted line is the function's average height or value. Although the function dips below and shoots above f(c), that's how high f(x) is on average.

Finding Distance Traveled

One of the best applications of definite integrals is their ability to play well with position and velocity functions. Remember that derivatives measure a rate of change? Well, it turns out that definite integrals measure accumulated change. For example, if you are given a function which represents the rate of sales of the new must-have toy, the Super Fantastic Hula Hoop, then the definite integral gives you the actual number of hula hoops sold.

Most often, however, math teachers like to explore this property of integrals when it comes to the position function. Specifically, the definite integral of the velocity function of an object gives you the distance traveled by the object. A word of caution: You will most often be asked to find the total distance traveled by the object—not to find the total *displacement*. To calculate the total distance, you'll first have to determine where the object changes direction (using a wiggle graph) and then integrate the velocity separately on every interval that direction changes.

Talk the Talk

Here's the difference between total distance traveled and **total displacement.** Let's say at any hour t, I want to know, in miles, how far I am away from my favorite bright orange '70s-style easy chair that my wife hates. My initial position (i.e., $t = 0$ hours) is in the chair, so position equals 0. Two hours later, I am at work, 50 miles away from my chair, so $s(2) = 50$ miles. Once my workday and commute home are complete, I am back home in the chair, and $s(12) = 0$. I have traveled a total distance of 100 miles, counting my travel away from the chair and back again. However, my displacement is 0. Displacement is the total change in position counting only the beginning and ending position; if the object in question changes direction any time during that interval of time, displacement does not correctly reflect the total distance traveled.

Example 3: In the book *The Fellowship of the Ring* by J.R.R. Tolkien, a young hobbit named Frodo embarks on an epic, exciting, and hairy-footed adventure to destroy the One Ring in the fires of Mount Doom. Based on a little estimation and the book *The Journeys of Frodo: An Atlas of J.R.R. Tolkien's The Lord of the Rings* by Barbara Strachey, we can design equations modeling Frodo's journey. In fact, during the first four days of his journey (from Hobbiton to the home of Tom Bombadil), his velocity (in miles per day) is given by the equation:

$$v(t) = -15.5t^3 + 86.25t^2 - 117.25t + 48.75$$

For example, $v(2)$ gives his approximate velocity at the exact end of the second day. Find the total distance Frodo travels from $t = 0$ to $t = 4$.

Solution: Since we want to find the total distance traveled, we need to see if Frodo changes direction at any point, and actually starts to wander toward Hobbiton rather than away. This is not necessarily caused by poor hobbit navigation, but perhaps by hindrances such as the old forest, getting caught in trees, etc. To see if his direction changes, do a wiggle graph for velocity (see Figure 16.6).

Figure 16.6

The hobbits have a pretty good sense of direction; in fact, they are heading farther and farther away from Hobbiton until just before the end of the fourth day.

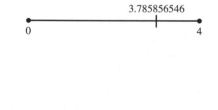

We will integrate the velocity equation separately, on both of these intervals. Because they are heading slightly backwards (i.e., towards their beginning point) on the interval (3.785856546,4), that definite integral will be negative. However, since it still represents the distance the hobbits are traveling, we don't want it to be subtracted from our answer, so we'll turn it into its opposite by multiplying it by –1. You should do this for any negative pieces of your wiggle graph in this type of problem. Therefore, the distance traveled will be …

$$\int_0^{3.785856546}\left(-15.5t^3 + 86.25t^2 - 117.25t + 48.75\right)dx - \int_{3.785856546}^4\left(-15.5t^3 + 86.25t^2 - 117.25t + 48.75\right)dx$$

Even though the numbers are darn ugly, the premise is very simple. I'll leave the figuring up to you. I get 108.2979533 for the first interval (distance away from Hobbiton) and –(–3.297953347) for the second, which is the small distance back towards Hobbiton; the sum is 111.596 miles.

Critical Point _____

Right about now, you're seeing that the numbers in this problem are not easy, whole numbers. They rarely turn out to be so in real-world (or Middle Earth) examples. Since calculus reform (a new movement among many mathematicians to make calculus more realistic and to embrace electronic calculation technology) tackles problems such as these, I wanted to throw in one or two problems that are much simpler if you use technology. There are those who would have me burned at the mathematical stake for suggesting such a thing. In fact, I was once yelled at fiercely by the lunch ladies in the high school cafeteria where I worked for suggesting that one use calculators to help check one's answers when converting fractions to decimals. I've never received fewer tater tots than I did that day. Lunch ladies can be so bitter.

You've Got Problems

Problem 3: When satellites circle closely around a planet or moon, the gravitational field surrounding the celestial body both increases the satellite's velocity and changes its direction in an orbital move called a "slingshot." (As you may know from the movie *Apollo 13*, Tom Hanks and his crew executed a slingshot maneuver around the moon to hurl themselves back toward earth.) Let's say that a ship executing this maneuver has position equation $s(t) = t^3 - 2t^2 - 4t + 12$, where t is in hours and $s(t)$ represents thousands of miles from earth. What is the total distance traveled by the craft during the first five hours?

Accumulation Functions

Before we close out this chapter and make it a fond memory, let's talk about accumulation functions. You'll probably see a few of them lurking around contemporary calculus classes, as they are now "in" since the advent of calculus reform. An *accumulation function* is a definite integral with a variable expression in one or more of its limits of integration. They are called accumulation functions because they get their value by accumulating area beneath curves, as do all definite integrals.

Practically speaking, you should be able to evaluate and differentiate these functions, so let's get to it. Believe it or not, evaluating accumulation functions is just as easy as evaluating any function—just plug in the correct x-value. Once you plug in the value, you apply the Fundamental Theorem to the resulting definite integral. For example, if given the function $f(x) = \int_{2}^{x}(t - 4)dt$

Talk the Talk _____

An **accumulation function** is a function defined by a definite integral; the function will have a variable in one or both of its limits of integration.

The most famous accumulation function is $\int_1^x \frac{1}{t}\,dt = \ln x$.
The natural log function gets its value by accumulating area under a very simple curve! For example, the value of ln 5, which always seemed so alien to me (where the heck do you get 1.60944?) is equal to the area beneath $y = \frac{1}{t}$ on the interval [1,5].

and asked to find $f(4)$, you would plug in 4 for x—not t, since this is a function of x, as denoted by $f(x)$—thereby making the upper limit of integration 4; then integrate as normal:

$$f(4) = \int_2^4 (t - 4)dx$$

$$= \left(\frac{t^2}{2} - 4t\right)\Big|_2^4$$

$$= (8 - 16) - (2 - 8) = -2$$

To find the derivative of an accumulation function, look no further than part two of the Fundamental Theorem.

For example, if $f(x) = \int_2^x (t - 4)dt$, then $f'(x) = x - 4$; just plug the top bound into t and multiply by its derivative (which is 1 in this case). Pretty easy, eh? You already knew how to tackle these problems, even before they showed up. Kudos!

You've Got Problems

Problem 4: If $g(x) = \int_{-\pi}^{x/2} \cos 2t \, dt$, evaluate

    (a) $g(4\pi)$

    (b) $g'(4\pi)$

The Least You Need to Know

- To find the area between two curves, integrate the curve on top subtracted by the curve below it on the proper interval.
- The average value of a function, $f(x)$, over an interval, $[a,b]$, is found by dividing the definite integral of that function, $\int_a^b f(x)dx$, by the length of the interval itself, $b - a$.
- To calculate the distance traveled by an object, calculate the definite integral of its velocity function separately for each period of time it changes direction.
- Accumulation functions get their value by gathering area under a curve; they are defined by definite integrals possessing variables in one or more of their limits of integration.

Chapter 17

Integration Tips for Fractions

In This Chapter

◆ Rewriting fractions to integrate

◆ Upgraded *u*-substitution

◆ Inverse trig functions as antiderivatives

◆ Integrating by completing the square

Let's be honest with each other for a moment. Integration sort of stinks. In fact, integration really stinks. It is much harder than taking derivatives (it's not just your imagination), because there is rarely a clear-cut way to integrate an expression if you can't use the Power Rule. With derivatives, if you have a fraction, you apply the Quotient Rule—end of story. No one's confused; everyone's happy. It's not the same with integrals.

If you're given an expression that can't be integrated with the Power Rule or basic *u*-substitution, there's a choice to be made. That choice is: What the heck do I try next? Although the vast majority of calculus problems you'll encounter can be integrated by one of those first methods, there is a plethora, a smorgasbord, a buffet of alternate techniques to finding integrals. Though none are really hard, none are as clear-cut as differentiation.

When integrating a fraction, there is no hard-and-fast rule that is applied every time. In fact, the methods available for integrating fractions are nearly endless. However, this chapter will add three of the most popular tools used to integrate fractions to your integral toolbox. If you're trying to drive a nail, you wouldn't use a screwdriver, would you? Well, if *u*-substitution is the metaphoric equivalent of an integration screwdriver, here come the integration chisel, socket wrench, and needle-nose pliers.

Separation

Breaking up is hard to do, but under specific circumstances, it is really quite worthwhile. Sometimes things just don't work out, and fractions have to go their separate ways. After a long, sunny time in the numerator together, terms just want a little more "me" time and some personal space. However, after all the time they've spent together, they've saved up a little bundle in the denominator, and both want to walk away with it. The good news is, in the math world, both pieces of the numerator get a full share of the denominator—no lawyers, no haggling over how it should be broken up. Both terms of the numerator walk away with a full denominator, and are a little wiser for having gotten involved in the first place.

Critical Point

Back in grade school, you learned that two fractions couldn't be added unless they had the same denominator. With this knowledge, you proudly calculated things like $\frac{1}{3} + \frac{7}{3} = \frac{7+1}{3} = \frac{8}{3}$, and never looked back. Well, look at it backward for just a moment. If you are given the fraction $\frac{a+b}{c}$, you can rewrite it as $\frac{a}{c} + \frac{b}{c}$, just as you know that $\frac{8}{3} = \frac{7}{3} + \frac{1}{3}$.

Top-heavy integrals (i.e., lots of terms in the numerator but only one in the denominator) and other fractional integrals are occasionally easier to solve if you split the larger integral into smaller, more manageable ones. Although the original problem couldn't be solved via *u*-substitution or the Power Rule, the smaller integrals usually can.

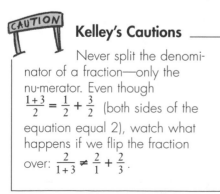

Kelley's Cautions

Never split the denominator of a fraction—only the nu-merator. Even though $\frac{1+3}{2} = \frac{1}{2} + \frac{3}{2}$ (both sides of the equation equal 2), watch what happens if we flip the fraction over: $\frac{2}{1+3} \neq \frac{2}{1} + \frac{2}{3}$.

Example 1: Find $\int \frac{x^4 - 2x^3 + 5x^2 - 3x + 1}{x^2}$ using the separation technique.

Solution: This is a fraction, so the Power Rule for Integration doesn't apply, and setting the numerator or denominator equal to *u* is not going to do a whole lot for you, so *u*-substitution is out. If, however, you separate the five terms of the large numerator into five separate fractions, watch what happens:

$$\int \frac{x^4}{x^2} \, dx - 2\int \frac{x^3}{x^2} \, dx + 5\int \frac{x^2}{x^2} \, dx - 3\int \frac{x}{x^2} \, dx + \int \frac{1}{x^2} \, dx$$

When you simplify each of these fractions, you get simple integrals, each of which can be integrated via the Power Rule for Integration:

$$\int x^2 \, dx - 2\int x \, dx + 5\int dx - 3\int \frac{1}{x} \, dx + \int x^{-2} \, dx$$

$$= \frac{x^3}{3} - 2\cdot\frac{x^2}{2} + 5\cdot\frac{x^1}{1} - 3\ln|x| + \frac{x^{-1}}{-1} + C$$

$$= \frac{x^3}{3} - x^2 + 5x - 3\ln|x| - \frac{1}{x} + C$$

You've Got Problems

Problem 1: Find $\int \frac{\sin x + \cos x}{\cos x} dx$ using the separation technique.

Tricky *U*-Substitution and Long Division

When we first discussed *u*-substitution, I made it a point to say that the derivative of *u* must appear in the problem. This is usually true, so I wasn't technically lying. There is a way to use *u*-substitution, even if it's not the most obvious choice.

Example 2: Find $\int \frac{2x-1}{x-2} dx$.

Solution: For grins, let's go ahead and try to find the antiderivative using *u*-substitution. Once again, remember our tip: If you're not sure what to set equal to *u*, try the denominator. Therefore, $u = x - 2$ and $du = dx$. If you make the appropriate substitutions back into the problem, you get:

$$\int \frac{2x-1}{u} \, du$$

To be honest, it doesn't look much better than the original, does it? Don't give up, though; we're not out of options. Go back to your *u* equation and solve it for *x* to get:

$$u = x - 2$$
$$x = u + 2$$

Now substitute that *x*-value into the numerator of our integral, and suddenly everything is a little cheerier:

$$\int \frac{2(u+2)-1}{u}\,du$$

$$= \int \frac{2u+3}{u}\,du$$

At least all of our variables are the same now. That's a relief. Can you see where to go from here? This fraction is top-heavy, with lots of terms in the numerator but only one in the denominator, so we can use the separation method from the last section to finish. What a happy coincidence that we just learned it!

$$\int \frac{2u}{u}\,du + \int \frac{3}{u}\,du$$

$$= 2\int du + 3\int \frac{1}{u}\,du$$

$$= 2u + 3\ln|u| + C$$

$$= 2(x-2) + 3\ln|x-2| + C$$

$$= 2x + 3\ln|x-2| + C$$

Critical Point

Unfortunately, space doesn't allow for a review of polynomial long division. If you can't remember how to do it, check out one of these web pages: www.purplemath.com/modules/polydiv.htm, www.sosmath.com/algebra/factor/fac01/fac01.html, and www.karlscalculus.org/notes.html.

You may be wondering why the –4 vanished in the last step of Example 2. Remember that C is some constant you don't know. If you subtract 4 from that, you'll get some other number (which is 4 less than the original mystery number). Since I *still* don't know the value for C, I just write it as C again, instead of writing $C-4$.

There are alternatives when integrating fractions like these. In fact, you can begin a rational integral by applying long division; it helps to simplify the problem *if the numerator's degree is greater than or equal to the denominator's degree*. It works like a charm if the denominator is not a single term, as is the case with Example 2.

Since the degree of the numerator (1) is greater than or equal to the degree of the denominator (1), begin by dividing $2x - 1$ by $x - 2$:

$$\frac{2x-1}{x-2} = x-2{\overline{)\,2x-1\,}}^{\,2} = 2 + \frac{3}{x-2}$$
$$\underline{-2x+4}$$
$$3$$

Therefore, we can rewrite the integral as $\int \left(2 + \frac{3}{x-2}\right)dx$, and tricky u-substitution is no longer required. The solution will again be $2x + 3\ln|x - 2| + C$.

You've Got Problems

Problem 2: Find $\int \frac{2x+1}{2x-3}dx$ using tricky u-substitution or by long division.

Integrating with Inverse Trig Functions

Let me preface this section with a bit of perspective. You are not going to have to integrate inverse trigonometric functions very often. In fact, you will probably use the formulas we're about to learn less often than the average male college student does laundry during one semester, and that's not a very big number, let me tell you. However, every once in a blue moon, you'll see a telltale sign in an integral problem—a particular fingerprint that tells you inverse trig functions will be part of your solution.

Way back when you were memorizing trig derivatives in Chapter 9, I threw the inverse function derivatives in at the end of the list. Hopefully you did, indeed, memorize them, or at least can pick them out of a lineup of the usual suspects. As we have done in previous sections, we are going to now progress from the derivative back to the original inverse trig function. For example, since $\frac{d}{dx}(\arctan x) = \frac{1}{1+x^2}$, we automatically know that $\int \frac{1}{1+x^2} dx = \arctan x + C$.

Don't think that we're limited to only that formula, however. The denominator does not have to be $1 + x^2$ in order to use the arctangent formula. Any constant can take the place of 1, and just about any function can replace the x^2. They're going to seem a little confusing, but I need to show you the formulas we'll be using before we get into examples. The examples will make everything a lot clearer; trust me.

In each of these formulas, a represents a constant and u represents a function (I use a u to represent the function to remind you that you'll have to do u-substitution for all of these inverse trig integral problems):

- $\int \frac{1}{\sqrt{a^2 - u^2}} du = \arcsin\left(\frac{u}{a}\right) + C$

- $\int \frac{1}{a^2 + u^2} du = \frac{1}{a}\arctan\left(\frac{u}{a}\right) + C$

- $\int \frac{1}{u\sqrt{u^2 - a^2}} du = \frac{1}{a}\text{arcsec}\left(\frac{|u|}{a}\right) + C$

These formulas represent identification patterns. If you see an integral with a square root in the denominator, and inside that square root is a number minus a squared function, alarm bells should go off in your head. This pattern matches the first formula, and may be integrated using the arcsine formula. If instead you have a denominator with a number plus a function squared but no square root, arctangent may be the way to go. Remember, there is no easy way to tell what integration method to use every time, so don't be discouraged if your first attempt doesn't work—just try another technique we've learned until something sticks.

> **CAUTION**
>
> **Kelley's Cautions**
>
> Pay attention to the order of subtraction if a radical appears in the denominator. In arcsine, you'll have a constant minus the function, whereas the order is reversed in arcsecant.

Example 3: Evaluate $\int \dfrac{\sin x}{\sqrt{4 - \cos^2 x}}\, dx$.

Solution: Integrating to get arcsine is a good option; you can tell because of two major indicators: (1) a square root containing subtraction in the denominator, and (2) the subtraction follows the order "constant minus function squared," suggesting the arcsine rather than the arcsecant function, whose subtraction order is reversed. Match this problem up with our new arcsine formula, ignoring the numerator for now. Because 4 must be the a^2 term, $a = 2$; similarly, $u = \cos x$.

As I mentioned earlier, the presence of the function u reminds you to do u-substitution, so $du = -\sin x\, dx$ and $-du = \sin x\, dx$. Using these three statements, we can rewrite the integral:

$$-\int \frac{du}{\sqrt{a^2 - u^2}}$$

This now exactly matches the formula for arcsine (the coefficient of -1 doesn't affect the contents of the integral, which are a perfect match), so we know:

$$-\int \frac{du}{\sqrt{a^2 - u^2}} = -\arcsin\left(\frac{u}{a}\right) + C$$

$$= -\arcsin\left(\frac{\cos x}{2}\right) + C$$

You've Got Problems

Problem 3: Find $\int \dfrac{3x}{7 + x^4}\, dx$ by integrating to get an inverse trig function.

Completing the Square

Most people hate fractions, even if they don't remember why anymore. I have finally graduated to only hating fractions of fractions. Did you ever notice that all gasoline prices are fractions of fractions? The big sign may say gas is \$1.25 a gallon, but if you look closer, you'll see that it reads \$1.25 and $\frac{9}{10}$ of a cent. But that is neither here nor there.

Integration might be getting on your nerves almost as much as fractions. The fabulous combination of integrating fractions is stressful enough to cause your eye to twitch and that voice in your head to take a different tone. If your patience with integration really started to thin out when you read about inverse trig functions, then completing the square at the same time you're integrating may just push you over the edge.

I have good news. The integration methods we'll learn in the next chapter are completely new, and actually kind of fun. To be honest, I think the methods we've learned to integrate fractions in this chapter are kind of a drag. But, as they say, it's not a personality contest, so we might as well grin and bear it through one more topic. Before you read on, make sure to review the section on completing the square from Chapter 2, and reread those formulas from the last section we did on inverse trig integration.

Integrating by completing the square is a useful technique in one specific circumstance—if you have a quadratic polynomial in the denominator and no variable at all in the numerator. The process itself is nothing more than a combination of things you've already done, but the process is not obvious to most people (including me) when you first see it, so it's worth working through an example.

Example 4: Find $\int \frac{4dx}{3x^2-6x+30}$ by completing the square.

Solution: In order to complete the square in the denominator, the coefficient of the x^2 term must be 1, so factor a 3 out of the denominator. While you're at it, go ahead and pull that 4 out of the top.

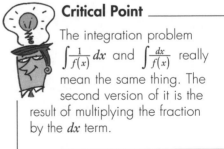

Critical Point

The integration problem $\int \frac{1}{f(x)}\,dx$ and $\int \frac{dx}{f(x)}$ really mean the same thing. The second version of it is the result of multiplying the fraction by the *dx* term.

$$\frac{4}{3} \cdot \int \frac{dx}{x^2 - 2x + 10}$$

Now let's focus on the denominator. According to our completing the square procedure from Chapter 2, we take half of –2 and square it to get 1. To avoid changing the value of the fraction, we must both add and subtract 1 in the denominator:

$$\frac{4}{3} \cdot \int \frac{dx}{x^2 - 2x + 1 - 1 + 10}$$

Now factor $x^2 - 2x + 1$ and combine $-1 + 10$ to get:

$$\frac{4}{3} \cdot \int \frac{dx}{(x-1)^2 + 9}$$

Notice that this is a fraction with a squared function being added to a constant in the denominator. Sound familiar? We can apply the arctangent integral formula from the last section, with $u = x - 1$ and $a = 3$. Remember to do u-substitution; in this case, $du = dx$, so no adjustments to du are necessary:

$$\frac{4}{3} \cdot \int \frac{du}{u^2 + a^2}$$

$$= \frac{4}{3} \cdot \frac{1}{3} \arctan\left(\frac{x-1}{3}\right) + C$$

$$= \frac{4}{9} \arctan\left(\frac{x-1}{3}\right) + C$$

You've Got Problems

Problem 4: Find $\int \frac{2}{(x-3)\sqrt{x^2 - 6x + 5}} dx$ by completing the square.

Selecting the Correct Method

Trying to decide which technique to use when integrating fractions can feel like trying to build a boat after it's already started to flood. The longer it takes, the less motivated you are to try and finish the job. Just remember that it takes time and practice. The smallest difference between integrals completely changes the way you approach the problem.

For example, consider the integral $\int \frac{dx}{x^2 + 16}$. The best way to solve this is via the arctangent formula with $u = x$ and $a = 4$. However, consider the integral $\int \frac{x}{x^2 + 16} dx$. Here, completing the square won't work! In fact, u-substitution is the best solution technique. It's hard to believe that simply adding one little "x" completely changes our approach. How about $\int \frac{x^2}{x^2 + 16} dx$? Now only long division will get you to the correct answer, since the degree of the numerator is greater than or equal to that of the denominator.

When push comes to shove with fraction integration, don't panic. Realize going into the problem that you may have to integrate a couple of times using a couple of different methods before it works out correctly. However, you should be able to tackle just about any elementary integral fraction now, as long as you're patient.

The Least You Need to Know

◆ You can split top-heavy fraction integral problems into separate fractions in order to create smaller, simpler integral problems.

◆ Sometimes tricky *u*-substitution is useful for integrating rational functions, especially if both the numerator and denominator are linear (i.e., have degree 1).

◆ If the numerator's degree is greater than or equal to the degree of the denominator, you can simplify an integral by applying polynomial long division before you integrate.

◆ If the denominator of the integral is a quadratic function and the numerator does not contain a variable, you can complete the square in the denominator, and you may be able to integrate the result using an inverse trigonometric function.

Chapter 18

Advanced Integration Methods

In This Chapter

◆ Integration by parts

◆ Applying the parts table

◆ Another way to integrate fractions

◆ Teaching improper integrals some manners

Without a doubt, my students always hated the integration methods in Chapter 17. It was probably their least favorite section, even though they admitted it wasn't the hardest. The biggest complaint they expressed was how similar the problems were, but how different the processes were that solved them. It didn't help that those methods aren't used that often, since it afforded hardly any opportunity to practice later in the course.

All this dislike of those topics actually helped them better understand the things we'll learn in this chapter. Integration by parts and by partial fractions are two of the most interesting things in calculus. For a reason that I can't explain, what you're going to learn about integration now makes the older stuff look like drudgery. Among the things you'll learn is one of the greatest (and most secret) calculus tricks of all time. While some instructors will mention it in passing, few willingly explain it in class.

There's something about math teachers that impels them to do things the longest and most difficult way possible. I've never quite decided if that's a good or bad thing, to be honest. In the words of another of my college professors, "Sure, math is hard, but why is hard bad?" (Did I mention that he was Austrian, so much of what he said, no matter how benign, made him sound like an evil supervillain? "Ah, Michael, you seem to have forgotten your constant of integration; I will have to remove your eyes and eat them in front of you ….") The thing is, math doesn't *have* to be so hard. Not when you know some of the tricks.

Integration by Parts

Even with all the integration methods we know so far, you're still going to encounter integrals you can't solve. It's a frustrating feeling, like trying to swim out of quicksand.

The more you struggle, the more you realize that your struggling isn't getting you anywhere. Well, the best defense against quicksand (according to the *Worst-Case Scenario Survival Handbook*) is a "stout pole," which you can stretch across the sand's surface and use to regain some leverage. Integrating by parts is your stout pole in the face of disgusting integration problems; you'll likely use it often, and you'll grow to depend on it. In fact, it'll become so handy, you'll probably end up using it to integrate things that could be done much more simply. However, pole in hand, you won't care, as you march confidently into the sunset. (*Curtain drops. Orchestra plays fanfare. House lights come up to thunderous applause.*)

Talk the Talk

Integration by parts allows you to rewrite the integral $\int u \, dv$ (where u is an easily differentiated function and dv is one easily integrated) as $uv - \int v \, du$. I call this formula the "brute force" method, because the tabular method you'll learn soon, by comparison, is much slicker.

The Brute Force Method

Your goal in *integration by parts* is to split the integral's contents into two pieces (including the dx but not the integral sign itself). One of the pieces (which we'll call u) should be easy to differentiate. The other piece (which we'll call dv) should be easy to integrate. Sometimes this process will require a little experimentation. Next, you'll differentiate u to get du (just like in u-substitution), and integrate dv to get v. Finally, plug all of those things into the integration by parts formula: $\int u \, dv = uv - \int v \, du$.

On the left side will be your original integral, so the right side represents a different (yet equivalent) way to rewrite it. It's sort of like the long division method you've already learned—this step allows you to rewrite the integral in a much more manageable form.

Critical Point

The integration by parts formula actually comes from the Product Rule. Let's say you have two functions, called $u(x)$ and $v(x)$. According to the Product Rule, $\frac{d}{dx}(u \cdot v) = u\,dv + v\,du$. If you integrate both sides of the equation, you get

$uv = \int u\,dv + \int v\,du$, and solving that equation for $\int u\,dv$ gives you the parts formula. Of course, all of the necessary "+C" terms are left out—you'll have to pop one of those on at the end if you're dealing with an indefinite integral.

Example 1: Find the solution to $\int x \sin x\,dx$ using integration by parts.

Solution: Either of these two pieces is easily integrated by itself, but together, they're trouble. You cannot rewrite the integral $\int x \sin x\,dx$ as $\int x\,dx \cdot \int \sin x\,dx$—only addition and subtraction inside the integral can be split up into separate integrals.

Even u-substitution is a bust. Of course, none of our fraction methods will work.

Time for parts—I will set $u = x$ and $dv = \sin x\,dx$. Even though I could have integrated the x and differentiated sine, it's usually best to pick a u that gets *less* complicated when you take the derivative. Therefore, differentiate u to get $du = dx$ and integrate the dv equation to get $v = -\cos x$. (Don't worry about the +C for now—we'll stick that in at the end.) Substitute these into the parts formula and solve the simple integral that results:

$$\int u\,dv = uv - \int v\,du$$
$$\int x \sin x\,dx = x(-\cos x) - \int(-\cos x)dx$$
$$= -x\cos x + \int \cos x\,dx$$
$$= -x\cos x + \sin x + C$$

You've Got Problems

Problem 1: Find the solution to $\int x^2 e^x\,dx$; you'll have to apply integration by parts twice, because the integral that results from the parts method can only be integrated by using parts again!

I promised you before that I'd tell you how to integrate the natural log function using parts. Set $u = \ln x$ and $dv = dx$. (Your dv term should always include the dx, even if that's all it includes.) Therefore, $du = \frac{1}{x}\,dx$ and $v = x$. According to the parts formula, $\int \ln x\,dx$ can be rewritten as $x \ln x - \int x \cdot \frac{1}{x}\,dx$, so integrate:

$$\int \ln x\,dx = x \ln x - \int dx$$
$$= x \ln x - x + C$$

The Tabular Method

There is a much easier way to do integration by parts problems, but a word of caution must precede it. This method will not work for all integration by parts problems. If the term you pick to be u does not eventually equal 0 when you take multiple derivatives, it will not work. In this procedure, you still begin by deciding upon the u and dv pieces the same way. Then, you draw a table with three columns, labeled "u," "dv," and "± 1."

> **CAUTION**
>
> **Kelley's Cautions**
>
> Don't forget to slap on a "$+C$" when you're integrating by parts, if the original integral is indefinite.

In the first row of the table, you write the u term in the first column, the dv term in the second, and a +1 in the third. In the second row, you'll write the derivative of the u term, the integral of the dv term, and this time a −1 in the ±1 column. In the third row, you take yet another derivative, another integral, and switch back to a +1.

So, the u column contains successive derivatives—$f(x)$, $f'(x)$, $f''(x)$, etc.—the dv column contains successive integrals $(g(x), \int g(x)\,dx, \int(\int g(x)\,dx)dx)$, and the final column begins with a +1 and alternates between −1 and +1 after that. Eventually, your u column will become 0. The first row that contains a 0 for u will be the last row in your table, except for one additional +1 or −1 in its own row. Figure 18.1 is the table generated by Example 1 $(\int x \sin x\,dx)$.

Figure 18.1

The first row contains the original u *and* dv *statements, and the next rows contain the derivatives and integrals, respectively. Notice the little negative term sitting by itself in the last row. Follow the arrows to the answer.*

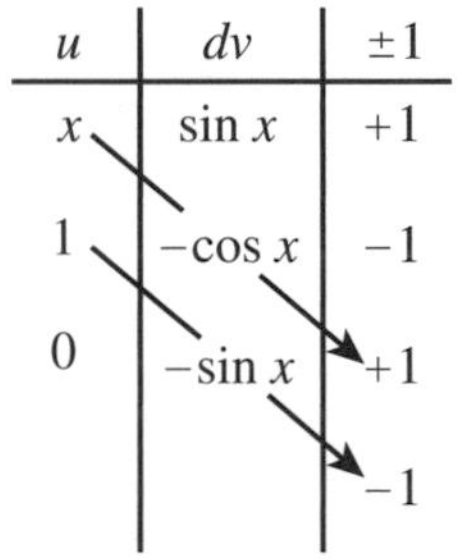

u	dv	± 1
x	$\sin x$	$+1$
1	$-\cos x$	-1
0	$-\sin x$	$+1$
		-1

Now draw a series of arrows that begin with the u term and go southeast from there, hitting the middle column of the second row and the third column of the third row. Multiply these things together to get one term of your answer, $x(-\cos x)(+1) = -x\cos x$. To get the second term of the answer, draw a similar arrow beginning with the second row: $1(-\sin x)(-1) = \sin x$. There is no third term, because a 0 stands in that position, so the final answer is $-x\cos x + \sin x + C$.

You've Got Problems

Problem 2: Redo the integral from Problem 1 $(\int x^2\, e^x\, dx)$ using the tabular method, and show that you get the same answer.

Integration by Partial Fractions

I hope there's still space left in your brain for one more fraction integration topic. Truth be told, it's actually kind of neat (the way that reading books like *Lord of the Flies* would be neat if your literature grade didn't depend on how well you understood the symbolism in it). Way back in algebra, you learned how to add rational expressions. For example, in order to add $\frac{5}{x+1}$ to $\frac{3}{x-5}$, you'd have to get a common denominator of $(x + 1)(x - 5)$ and simplify like so:

$$\frac{5}{x+1} \cdot \frac{x-5}{x-5} + \frac{3}{x-5} \cdot \frac{x+1}{x+1}$$
$$= \frac{5(x-5) + 3(x+1)}{(x+1)(x-5)}$$
$$= \frac{8x - 22}{(x+1)(x-5)}$$

Critical Point

You can tell how much time the tabular method saves you compared to the brute force method. If your table is longer than three full rows, then each additional row represents a time you'd have to reapply integration by parts in the same problem.

Now we're going to learn how to go from $\frac{8x-22}{(x+1)(x-5)}$ to $\frac{5}{x+1} + \frac{3}{x-5}$. The process is called *partial fraction decomposition*, and it is a great way to integrate a fraction whose denominator is factorable.

Example 2: Use partial fraction decomposition to find $\int \frac{x+2}{x^2(x-1)} dx$.

Solution: The denominator will not always be factored for you, but in this case it is. It is composed of two linear factors (x and $x - 1$), one of which is squared. In the process of partial fractions, this is called a *repeating factor*. We can rewrite $\frac{x+2}{x^2(x-1)}$ as a sum of three fractions whose denominators are individual factors of the original denominator:

$$\frac{x+2}{x^2(x-1)} = \frac{A}{x} + \frac{B}{x^2} + \frac{C}{x-1}$$

The A, B, and C are unknown constants. It is our goal to find them by solving that equation. To make things easier, multiply both sides of the equation by $x^2(x-1)$:

$$x + 2 = Ax(x - 1) + B(x - 1) + Cx^2$$
$$x + 2 = Ax^2 - Ax + Bx - B + Cx^2$$
$$x + 2 = x^2(A + C) + x(B - A) - B$$

Notice that I grouped the terms by factoring out common variables. This makes things easy. Remember, both sides of the equation must be equal, and since there is no x^2 term on the left, then there can't be one on the right, making $A + C = 0$. Also, the x term on the left has coefficient 1, so the same must be true on the right: $B - A = 1$. Similarly, $2 = -B$, so $B = -2$ (i.e., the constants on both sides must be equal). Since we know B, we can find A:

$$B - A = 1$$
$$(-2) - A = 1$$
$$-A = 3$$
$$A = -3$$

Talk the Talk

Partial fraction decomposition is a method of rewriting a fraction as a sum and difference of smaller fractions, whose denominators are factors of the original, larger denominator. If a factor in the denominator is raised to a power, it is called a **repeating factor**. The factor $(x + a)^n$ will show up n times in partial fraction decomposition, with ascending powers, beginning from 1. In other words, if the denominator contains the repeating factor $(x - 3)^5$, your decomposition will contain $\frac{A}{x-3}$, $\frac{B}{(x-3)^2}$, $\frac{C}{(x-3)^3}$, $\frac{D}{(x-3)^4}$, and $\frac{E}{(x-3)^5}$. Even though these powers get high, you still consider the factors linear, since there is still a degree of 1 inside the parentheses. In Example 2, the repeating factor is x, which is why x and x^2 show up in separate fractions.

Finally, because $A + C = 0$, $C = 3$. Now, plug these values into the partial fraction formula:

$$\frac{x + 2}{x^2(x - 1)} = \frac{A}{x} + \frac{B}{x^2} + \frac{C}{x - 1} = \frac{-3}{x} + \frac{-2}{x^2} + \frac{3}{x - 1}$$

The original fraction can be rewritten as three smaller fractions, each of which can be integrated using the natural log formula or the Power Rule for Integration:

$$\int \frac{x + 2}{x^2(x - 1)} dx = \int \frac{-3}{x} dx + \int \frac{-2}{x^2} dx + \int \frac{3}{x - 1} dx$$

$$= -3 \int \frac{1}{x} dx - 2 \int \frac{1}{x^2} dx + 3 \int \frac{1}{x - 1} dx$$

$$= -3 \ln|x| + \frac{2}{x} + 3 \ln|x - 1| + C$$

As you can see, the whole point of partial fraction decomposition is to break down the problem into manageable fractions. One word of caution: You only use numerators of A,

B, and C when the denominators are linear factors (repeated or not). The numerator of a partial fraction is always one degree less than the denominator, and a constant numerator (degree 0) with a linear denominator (degree 1) fits that description. If one of the factors in the denominator is a quadratic, then you will use a linear numerator (like $Ax + B$) for that partial fraction. For instance, here's a sample decomposition of a fraction containing a quadratic repeated factor:

$$\frac{3x - 5}{(x - 4)(x^2 + 1)^2} = \frac{A}{x - 4} + \frac{Bx + C}{(x^2 + 1)} + \frac{Dx + E}{(x^2 + 1)^2}$$

You've Got Problems

Problem 3: Find the solution to the integral $\int \frac{2}{x^2 + 4x + 3} dx$ by partial fraction decomposition.

Improper Integrals

There are lots of reasons that an integral can be termed *improper*. For example, integrals that make loud, sucking sounds through their teeth at the dinner table while loudly discussing their views on current events are most certainly improper. Not to mention integrals that tailgate you so closely on the road that you can almost read their upper limit of integration in your rearview mirror. However, we don't usually get to observe this sort of rude behavior from our mathematical symbols, so we must content ourselves with observing telltale mathematical signs that integrals are improper. The two most common indicators of an improper integral are infinite discontinuities and infinite limits of integration.

In essence, either the integration boundaries of the problem or one of the numbers between the boundaries causes trouble. In the integral problem $\int_2^\infty \frac{dx}{x}$, the impropriety is obvious. Sure, you can integrate the fraction to get $\ln|x|$, but how are you supposed to plug ∞ into that? Infinity is not a number!

Can you spot the problem in the integral: $\int_{-4}^0 \frac{5}{x + 4} dx$? Again, integration is no sweat (u-substitution and the natural log function again), but it doesn't make sense for -4 to be an integration boundary. The function $\frac{5}{x + 4}$ doesn't even exist at -4, so how can there be area under there?

Believe it or not, integrals that look like $\int_2^\infty \frac{dx}{x}$ and $\int_{-4}^0 \frac{5}{x + 4} dx$ sometimes actually represent a finite area (although neither of those two do). In order to solve them, however, we'll have to revisit an old, old friend. After all this time, limits have come back to haunt you, like that breakfast burrito from yesterday morning.

Example 3: Evaluate $\int_4^7 \frac{dx}{\sqrt{x-4}}$ even though it is an improper integral.

Solution: This integral is improper because its lower boundary (4) is a point of infinite discontinuity on the function being integrated. To remedy this, replace the troublesome boundary with the generic constant a, and allow a to approach 4 using a limit: $\lim\limits_{a \to 4+} \int_a^7 \frac{dx}{\sqrt{x-4}}$. (Technically, we have to approach 4 from the right, since the graph doesn't exist to the left of 4.)

Critical Point

In Example 3, the integral is improper because of one of its boundaries. You may also have problems with the values between an integral's boundaries. For example, $\int_{-2}^1 \frac{dx}{x^2}$ is not improper because of the boundaries of –2 or 1, but

because the function $\frac{1}{x^2}$ has an infinite discontinuity at $x = 0$, which falls between the boundaries. To solve a problem like that, you'd split the integral into two pieces, each piece of which would contain the troublesome boundary. For notation purposes only, we'll use a left-hand limit for the left piece of the integral and a right-hand limit for the right piece.

$$\int_{-2}^1 \frac{dx}{x^2} = \int_{-2}^0 \frac{dx}{x^2} + \int_0^1 \frac{dx}{x^2}$$
$$= \lim_{a \to 0} \left(\int_{-2}^a \frac{dx}{x^2} \right) + \lim_{a \to 0+} \left(\int_a^1 \frac{dx}{x^2} \right)$$

It might feel like we're cheating, but we're not. The lower limit is not 4 anymore—now it's a number insanely close to, but not quite, 4. Proceed with the integration using u-substitution with $u = x - 4$ and $du = dx$ (don't forget to plug the boundaries into the u equation to get u boundaries):

$$\lim_{a \to 4+} \int_{a-4}^3 u^{-1/2} du$$
$$= \lim_{a \to 4+} \left(2\sqrt{u} \Big|_{a-4}^3 \right)$$
$$= \lim_{a \to 4+} \left(2\sqrt{3} - 2\sqrt{a-4} \right)$$

We've ignored the limit long enough. To finish, we can use the substitution method from Chapter 6:

$$\lim_{a \to 4+} \left(2\sqrt{3} - 2\sqrt{a-4} \right) = 2\sqrt{3} - 2\sqrt{4-4} = 2\sqrt{3}$$

You've Got Problems
Problem 4: Evaluate $\int_{1}^{\infty} \dfrac{dx}{\left(\sqrt{x}\right)^{3}}$ even though it is an improper integral.

The Least You Need to Know

◆ Some difficult integrals can be integrated via integration by parts, where the *u* part represents an easily differentiated piece of the integral, and the *dv* part represents an easily integrated piece.

◆ The tabular method of integration by parts makes the technique much easier, but can only be applied if successive derivatives of *u* eventually become equal to 0.

◆ If the denominator of an integral is factorable, try to break up the integral using partial fraction decomposition; the smaller fractions will likely be much easier to solve.

◆ If a definite integral has an integration limit of ∞ or an infinite discontinuity on the interval over which you wish to integrate, it is called an improper integral, and must be solved using a limit that replaces the troublesome value.

Applications of Integration

In This Chapter

- ◆ Pump up the (rotational) volume
- ◆ Calculating holey volume, Batman!
- ◆ Pretending you're Noah: finding arc length
- ◆ Three-dimensional surface area

This chapter could really have been named many different things. I could have called it "Applications of the Fundamental Theorem," but there's already a chapter with that name, and a chapter title like "Revenge of Fundamental Theorem Applications" sounds too scary, and may have necessitated a parental warning sticker for the front of the book. This is a book on *calculus*, for goodness' sake, so anything that could hurt sales is definitely a no-go.

I also could have called it "Common Uses of the Definite Integral," but that sounds too much like an infomercial. It comes off sounding like a car salesman trying to convince you of something. Ever since the last car salesman I dealt with stared uncomprehendingly at me for a half-hour before confessing that he was mesmerized by what he called my "very large forehead," I am a little skittish about salesmen in general. So I opted for the plain, vanilla, unexciting title "Applications of Integration."

What you may not know (if you haven't immersed yourself in a study of the table of contents) is that this is it for integrals. Once you've finished this chapter, it's time to plunge into that brief but happy purgatory between topics, where things are strange, new, and slightly easier for a bit. Until then, it's time to memorize some new formulas.

Volumes of Rotational Solids

Have you ever seen one of those tissue paper accordion-style decorations? They start out as a two-dimensional cardboard shape, but as you open them, you get this cool-looking three-dimensional design. I see them most often in wedding receptions—little tissue stalactites hanging from hotel ballroom ceilings. For those of you who go to classier weddings than I, or have no idea what I'm referring to (these decorations are not as hot as they were during the '70s), Figure 19.1 can be used as a visual aid.

Figure 19.1

When a simple shape is rotated in three dimensions about an axis, it can create this wedding bell—a geometric reminder of your inability to commit.

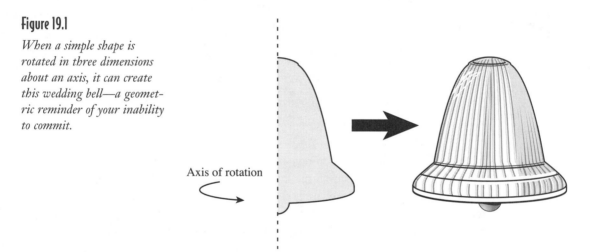

Axis of rotation

Our goal is to find the volume of shapes like these. We'll start with a function graph and rotate the area it captures around a horizontal or vertical axis, much like the area of the half-bell above is rotated around its vertical axis.

The Disk Method

As long as the rotational solid resulting from your graph has no hollow space in it, you can use the disk method to calculate its volume. You'll understand better what I mean by this later. The key to these problems is finding the *radius of rotation*, a length beginning at the rotational axis and extending to the outer edge of the area being rotated. In the disk method, the radius of rotation will *always* be perpendicular to the rotational axis.

It's very easy to locate and draw the radius of rotation, and doing so is an important step for any rotational solid problem. Once you find it, all you have to do is plug into the disk method formula: $V = \pi \int_a^b \left(r(x)\right)^2 dx$ (where a and b are the x-boundaries of the area you're rotating).

Talk the Talk

The **radius of rotation** is a line segment extending from the axis of rotation to the edge of the area being rotated. If this radius is vertical, it means that all of the functions you're dealing with *must* contain x-variables. On the other hand, if the radius of rotation is horizontal, you must use functions of y. The disk method is based on the fact that a rotational solid with no hollow spots will always have a circular cross section. In essence, you are integrating its cross-sectional area (which is πr^2, the area of a circle).

Example 1: Rotate the area bounded by $f(x) = -x^2 + 4$ and the x-axis about the x-axis, and calculate the volume that results.

If you can't visualize it, the solid resulting from the rotation described by Example 1 looks like Figure 19.2.

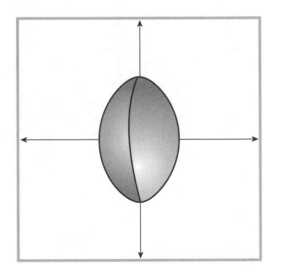

Figure 19.2

Don't worry—your grade won't be dependent upon how well you can draw three-dimensional figures.

Solution: Start by drawing everything and find the radius of rotation, as shown in Figure 19.3.

Figure 19.3

The radius of rotation extends from what you're rotating around to the edge of the region being rotated.

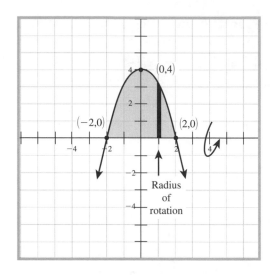

Critical Point _____

The equation of a semicircle, centered at the origin, of radius r is $y = \sqrt{r^2 - x^2}$.
If you rotate the region under the graph of this equation about the x-axis and treat r like a number (it's a constant, not a variable), you'll get the equation for the volume of a sphere: $V = \frac{4}{3}\pi r^3$. That's where that dang formula comes from.

The shaded region is the area that will be rotated about the x-axis. The dark line is radius of rotation—it extends from the axis of rotation (the x-axis) up to the edge of the graph f. Even though I've drawn the radius at $x = 1$, it will have the same defining formula anywhere on [–2,2]. Notice that the radius of rotation is perpendicular to the rotational axis, which is horizontal.

Find the length of the radius of rotation, $r(x)$, using the same reasoning you did when finding the area between curves: Its length is the top curve minus the bottom curve. Since the top curve is $f(x) = -x^2 + 4$ and the bottom curve is the x-axis ($g(x) = 0$), the radius of rotation is $r(x) = -x^2 + 4 - 0 = -x^2 + 4$. According to the disk method, the volume of the rotational solid will be:

$$V = \pi \int_a^b \left(r(x)\right)^2 dx$$
$$= \pi \int_{-2}^2 \left(-x^2 + 4\right)^2 dx$$
$$= \pi \int_{-2}^2 \left(x^4 - 8x^2 + 16\right) dx$$
$$= \frac{512\pi}{15}$$

Notice that the radius of rotation was vertical. This indicates that you had to use x-functions. Had it been horizontal, all of your functions would have to be in terms of y—i.e., $f(y)$ instead of $f(x)$. The only other difference with horizontal radii is that you find their length by subtracting the right function minus the left function, rather than top minus bottom.

You've Got Problems

Problem 1: Find the volume of the rotational solid generated by rotating the area in the first quadrant bounded by $y = x^2$, the y-axis, and the line $y = 9$ around the y-axis.

The Washer Method

If the solid of revolution has any hollow spots on it, you'll have to use a modified version of the disk method, called the *washer method*. It gets its name from the fact that the cross sections look like washers (i.e., they have holes or hollow spaces in their middles). Come to think of it, this process not only is a lifesaver; it can also find the volume of one!

Just like the disk method, the radius of rotation in the washer method must be perpendicular to the axis of rotation. Again, the orientation of that radius will tell you what variables to use in your calculations (vertical means x's and horizontal means y's). The only difference is that in the new method, you'll actually have *two* radii of rotation. One radius stretches from the axis of rotation to the outside edge of the area being rotated (just like before); this is called the *outer radius*. The other radius of rotation again originates from the rotational axis, but stretches to the inner edge of the area being rotated; this is called the *inner radius*. The formula for washer method subtracts the square of the inner radius from the square of the outer radius:

$$V = \pi \int_a^b \left(\left(R(x) \right)^2 - \left(r(x) \right)^2 \right) dx$$

Talk the Talk

The **washer method** is used to calculate the volume of a rotational solid even if part of it is hollow. Essential to the process are the **outer** and **inner radii** of rotation, which extend from the axis of rotation, respectively, to the far and near edges of the region to be rotated.

Critical Point

In essence, the washer method is the disk method repeated. First, it calculates the volume of the solid, ignoring the hole $\left(\pi \int_a^b \left(R(x) \right)^2 dx \right)$, finds the volume of the hole itself $\left(\pi \int_a^b \left(r(x) \right)^2 dx \right)$, and then subtracts the hole's volume from the overall volume.

Example 2: Consider the area captured between the graphs of $y = -x^2 + 2x + 1$ and $y = 1$. What volume is generated if this area is rotated about the x-axis?

Solution: Let's have a look at the situation in Figure 19.4.

Figure 19.4

Rotating this area about the x-axis results in something that is not completely solid— it has a hole of radius 1. It's almost a math Cheerio.

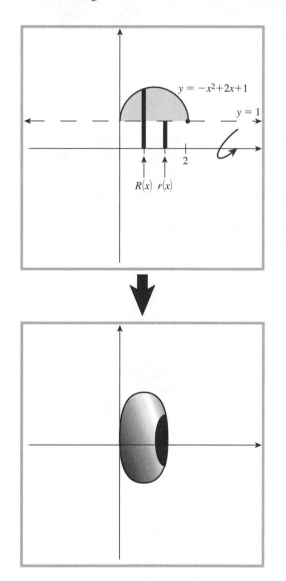

Because the shaded area is not right up against the x-axis, that space separating the region from the rotational axis also gets rotated, leaving a hole of radius 1 in the rotational solid. The outer radius, designated in the figure as $R(x)$, extends from the x-axis to the furthest edge of the region, whereas the inner radius, $r(x)$, extends to the innermost edge of the region. Find the lengths of these radii as we did with the disk method; subtract the top boundary equation minus the bottom boundary. Keep in mind that the bottom boundary is the *x*-axis and has the equation *y* = 0:

$$R(x) = -x^2 + 2x + 1 - 0 = -x^2 + 2x + 1$$
$$r(x) = 1 - 0 = 1$$

Finally, to find the area, plug into the washer method formula. Since these graphs intersect at *x* = 0 and *x* = 2, these are the boundaries for the region and should be your limits of integration:

$$\pi \int_0^2 \left(\left(-x^2 + 2x + 1 \right)^2 - (1)^2 \right) dx$$
$$= \pi \int_0^2 \left(x^4 - 4x^3 + 2x^2 + 4x \right) dx$$
$$= \frac{56\pi}{15}$$

A brief word of warning: We've been rotating about the *x*- and *y*-axes exclusively. This makes it easy to find the lengths of the radii of rotation, as the bottom boundary is usually 0, and subtracting 0 is, arguably, even easier than you might think. In Problem 2, you'll be rotating around something other than an axis, so be careful when calculating the radii of rotation.

You've Got Problems

Problem 2: Find the volume generated by rotating the region bounded by $y = \sqrt{x}$ and $y = x^3$ about the line $y = -1$.

The Shell Method

The final technique for finding rotational volumes uses radii *parallel* to the axis of rotation, rather than perpendicular to it. It's easy to remember that because *shell* and *parallel* rhyme. The other major difference in the shell method is the use of a *representative radius* rather than radii of rotation. Instead of extending a radius from the axis of rotation to the edge of the region, simply extend a radius from one edge of the region to the other. The formula for the shell method is $V = 2\pi \int_a^b d(x) \cdot h(x) dx$, where *d*(*x*) is the distance

Talk the Talk

In the shell method, you use a **representative radius,** which extends from one edge of the region to the opposite edge, rather than a radius of rotation, which extends from the axis of rotation to an edge of the region.

from your representative radius to the rotational axis, $h(x)$ is the length of the radius, and a and b are the boundaries of the area to be rotated.

The shell method can be used to calculate the volume of a rotational solid whether or not it has any hollowness. Therefore, it can be used in lieu of both the disk and washer methods.

Example 3: Rotate the area bounded by $y = x^3 + x$, $x = 2$, and the x-axis around the y-axis, and calculate the volume of the solid of revolution.

Solution: Since we're rotating about the y-axis, the washer method would require us to use horizontal radii (they are perpendicular to the vertical axis of rotation). That means everything would have to be in terms of y, and there's no easy way to solve $y = x^3 + x$ for x. However, with the shell method you don't have to do any conversions at all. Draw the region with its representative radius—it should be a vertical segment running across the region since the axis of rotation is also vertical (see Figure 19.5).

Figure 19.5

Notice that the radius is now parallel to the rotational axis.

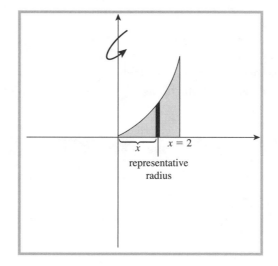

We'll say that the radius is a value of x units to the right of the origin, so $d(x) = x$. The height of the radius, $h(x)$, is found by subtracting its bottom boundary from its top boundary: $x^3 + x - 0$. Since we are using all x-values, the boundaries of the region are 0 and 2. Plug these values into the formula for the shell method:

$$2\pi \int_0^2 x \cdot \left(x^3 + x\right) dx$$
$$= 2\pi \int_0^2 \left(x^4 + x^2\right) dx$$
$$= \frac{272\pi}{15}$$

Keep in mind that the equation $d(x)$ will not always be x, as it is in this example. In fact, if we rotate the region about the line $x = -2$, we get Figure 19.6.

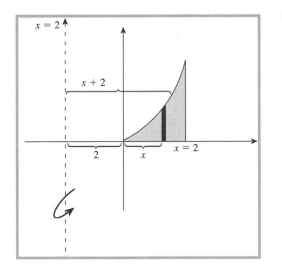

Figure 19.6

The region is the same as Example 3, but the axis of rotation is changed.

The length from the origin to the radius is still x, but there is an additional distance of 2 to the rotational axis, so $d(x) = x + 2$.

Arc Length

At this point, you can do all kinds of crazy math calculating. Geometry told you how to find weird areas, and calculus took that one step farther. The kinds of areas you can calculate now would have boggled your mind back in your days of geometric innocence. However, it remains a math skill that has visible and understandable application, even to those who don't know the difference between calculus and a tuna sandwich. Now, let's add to your list of skills the ability to find lengths of curves. By the time you're done, you'll even be able to prove (finally) that the circumference of a circle really is 2π.

Rectangular Equations

The term "rectangular equations" really means "plain, old, run-of-the-mill, everyday functions." Mathematicians use it for the obvious reason that it takes less time to say (mathematicians are busy people). It turns out that finding the length of a curve (on some x interval) is as easy as calculating a definite integral. In fact, the length of a continuous function $f(x)$ on the interval $[a,b]$ is equal to $\int_a^b \sqrt{1+\left(f'(x)\right)^2}\,dx$. In other words, find the derivative of the function, square it, add 1, and integrate the square root of the result over the correct interval.

Example 4: Find the length of the function $g(x) = \sqrt{x}$ between the points (1,1) and (16,4) on its graph.

Solution: Use the Power Rule to find the derivative of $g(x) = x^{1/2}$, and you get $g'(x) = \frac{1}{2\sqrt{x}}$. All you do now is plug into the arc length formula:

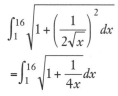

$$\int_1^{16} \sqrt{1+\left(\frac{1}{2\sqrt{x}}\right)^2}\,dx$$
$$=\int_1^{16} \sqrt{1+\frac{1}{4x}}\,dx$$

The integration problem that results is not simple at all. For our purposes, it is enough to know and apply the formula, not to struggle through the integral itself. You'll find that many (if not most) arc-length integrals will end up complicated and require somewhat advanced methods to integrate. We will, however, satisfy ourselves with a computer- or calculator-assisted solution—they have no problem with complex definite integrals. The final answer is approximately 15.3397.

Don't feel like you're cheating by using a calculating tool rather than solving this problem by hand. You'd have to know just about every integration technique there is to find the arc lengths of even very simple functions.

You've Got Problems

Problem 4: Which function is longer on the interval [0,2]: $f(x) = x^2$ or $g(x) = x^3$? Find the length of each and compare.

Parametric Equations

We haven't mentioned parametric equations for a while—they've been lurking in the shadows, but now they get to come out and play. There are numerous similarities in the formula for parametric equation arc length and rectangular arc length. Both are definite integrals, both involve a sum of two terms beneath a radical, and both involve finding the derivative of the original equation.

The arc length of a curve defined parametrically is found with the definite integral $\int_a^b \sqrt{\left(\frac{dx}{dt}\right)^2 + \left(\frac{dy}{dt}\right)^2}\, dt$, where a and b are limiting values of the parameter this time—not x boundaries. In other words, find the derivatives of the x and y equations, square them both, add them together, and square root and integrate the whole mess.

Example 5: The parametric representation of a circle with radius 1 (centered at the origin) is $x = \cos\theta$, $y = \sin\theta$. Prove that the circumference of a circle really is 2π by calculating the arc length of the parametric curve on $0 \le \theta \le 2\pi$.

Solution: Start by finding the derivatives of the x and y equations with respect to θ (it's easy):

$$\frac{dx}{d\theta} = -\sin\theta \text{ and } \frac{dy}{d\theta} = \cos\theta$$

Kelley's Cautions

Don't get confused because the parameter in Example 5 is not t. The formula for arc length with a parameter of θ is exactly the same; it just has θ's instead of t's in the formula.

Now, plug those values into the parametric arc-length formula and simplify using the Mama theorem (review Chapter 4, if you don't know what the heck I mean by that):

$$\int_0^{2\pi} \sqrt{\left(-\sin\theta\right)^2 + \left(\cos\theta\right)^2}\, d\theta$$
$$= \int_0^{2\pi} \sqrt{1}\, d\theta$$
$$= \left(\theta\right)\Big|_0^{2\pi}$$
$$= 2\pi - 0 = 2\pi$$

You've Got Problems

Problem 5: Find the arc length of the parametric curve defined by the equations $x = t + 1$, $y = t^2 - 3$ on the t interval $[1,3]$.

The Least You Need to Know

- The disk method is used to find the volumes of rotational solids as long as they contain no hollow parts. If there are hollow parts in the rotational solid, you must use the washer or shell methods.
- Both the disk and the washer method use rotational radii that are perpendicular to the axis of rotation, whereas the radii used in the shell method are parallel to the axis of rotation.
- The washer method is actually the disk method appearing twice in the same problem.
- You can find the arc length of rectangular or parametric curves via similar definite integral formulas.

Part 5

Differential Equations, Sequences, and Series

If differentiation and integration are, respectively, the mother and father figures of calculus, then sequences and series are the bratty kids. If you look closely, you can see that they're related. However, once you're knee-deep in sequences and series, you'll wonder how they can be related to their parents at all, and wonder if they were outright adopted. The closest you've come to dealing with infinite series and sequences is improper integration, but even that is a stretch.

Differential equations, on the other hand, are very obviously related to what we've been doing all along. In fact, solving them will begin as a simple extension of integration, something at which you are already approaching expert status. This process (called separation of variables) will not always work, however, so you'll also learn ways of dealing with differential equations that defy separation.

Differential Equations

In This Chapter

- ◆ What are differential equations?
- ◆ Separation of variables
- ◆ Initial conditions and differential equations
- ◆ Modeling exponential growth and decay

Most calculus courses contain some discussion of differential equations, but that discussion is extremely limited to the basics. Most math majors will tell you that they had to suffer through an entire course on solving differential equations at some point in their math career. This is because differential equations are extremely useful in modeling real-life scenarios, and are used extensively by scientists.

A differential equation is nothing more than an equation containing a derivative. In fact, you have created more than your fair share of differential equations simply by finding derivatives of functions in the first half of the book. In this chapter, we'll begin with the differential equation (i.e., the derivative) and work our way backwards to the original equation. Sound familiar? Basically we're just going to be applying integration methods, as we have for numerous chapters now.

However, solving differential equations is not the same thing as integrating. There are lots of more complex differential equations (that we won't be exploring) requiring much more complicated methods of solution. Luckily, the most popular differential equation application in beginning calculus (exponential growth and decay) requires us to use a very simple solution technique called separation of variables. Let's start there.

Separation of Variables

If a *differential equation* is nothing more than an equation containing a derivative, and solving a differential equation basically means finding the antiderivative, then what's so hard about solving differential equations, and why does it get treated as a separate topic? The reason is that differential equations are usually not as straightforward as this one:

$$\frac{dy}{dx} = 3\cos x + 1$$

Talk the Talk

Differential equations are just equations that contain a derivative. Most basic differential equations can be solved using a method called **separation of variables,** in which you move different variables to opposite sides of the equation so that you can integrate both sides of the equation separately.

Clearly, the solution to this differential equation is $y = 3\sin x + x + C$. All you have to do is integrate both sides of the equation. Most differential equations are all twisted up and knotted together with variables all over the place, like this:

$$\frac{dy}{e^{2x}} = xy\, dx$$

It looks like someone chewed up a whole bunch of equations and spat them out in random order (a puzzling and unappetizing description). In order to solve this differential equation, you'll have to separate the variables. In other words, move all the *y*'s to the left side of the equation and all the *x*'s to the right side. Once that is done, you'll be able to integrate both sides of the equation separately. This process, appropriately called *separation of variables*, solves any basic differential equations you'll encounter.

Example 1: Find the solution to the differential equation $\frac{dy}{dx} = ky$, where *k* is a constant.

Solution: We need to get that *y* to the left side of the equation and the *dx* to the right side. Since *k* is a constant, it's not clear whether or not I should move it. As a rule of thumb, move all constants to the right side of the equation. Our goal is to solve for *y*, so I don't want any non-*y* things on the left side of my equation. Let's start by moving the *y*; divide both sides of the equation by *y* to get:

$$\frac{dy}{y \cdot dx} = k$$

Now let's shoot that dx to the right side of the equation by multiplying both sides by dx:

$$\frac{dy}{y} = k\,dx$$

At this point, we can integrate both sides of the equation. Since k is a constant, its antiderivative will be kx (not $\frac{dy}{dx}$), just like the antiderivative of 5 would be $5x$:

$$\int \frac{dy}{y} = \int k\,dx$$

$$\ln|y| = kx + C$$

We're not quite done yet. Our final answer for a differential equation should be solved for y. To cancel out the natural log function and accomplish this, we have to use its inverse function, e^x, like this: $e^{\ln|y|} = e^{kx+C}$. (We'll drop the absolute value signs around the y now—they were only needed since the domain of the natural log function is only positive numbers. As the natural log disappears, we'll let the absolute value bars go with it.)

In other words, rewrite the equation so that both sides are the powers of the natural exponential function. This gives you $y = e^{kx+C}$, since e^x and $\ln x$ are inverse functions, and as such, $e^{\ln x} = \ln(e^x) = x$. We could stop here, but I want to go even one step further. Remember our basic exponential rules that said $x^a + x^b = x^{a+b}$? Our equation above looks like x^{a+b}, so I am going to break it up into $x^a \cdot x^b$:

$$y = e^{kx} \cdot e^{C}$$

Almost done. I promise. Since I have no idea what value C has (since it's the constant of integration), I have no idea what e^C will be. I know it'll be some number, but I have no idea what number that is. As we've done in the past, we'll rewrite e^C as C, signifying that even though it's not the same value as the original C, it's still some number we don't know: $y = Ce^{kx}$.

That's the solution to the differential equation. It took us a while to get here, but this is a very important equation, and we're going to need it in a few pages.

You've Got Problems

Problem 1: Find the solution to the differential equation $\left(x^2 - 1\right)dy = \frac{x\,dx}{\cos y}$.

Types of Solutions

Just like integrals, solutions to differential equations come in two forms: with and without a "$+C$" term. Definite integrals had no such term, because their final answers were numbers rather than equations. However, the solution to a differential equation will always come in delicious, equation form (with candy-shaped marshmallows). In some cases, though, you'll be able to determine exactly what the value of C should be, so you can provide a very specific answer.

Family of Solutions

If you are only given a differential equation, you can only get a general solution. Example 1 and Problem 1 are two such instances. Remember, integration cannot usually tell you exactly what a function's antiderivative is, since any equations differing only by a constant will have the same derivative.

The solution to a differential equation containing a "+C" term is actually a *family of solutions*, since it technically represents an infinite number of possible solutions to the differential equation. Think about the differential equation $\frac{dy}{dx} = 2x + 7$. If you use the separation of variables technique, you get a solution of $y = x^2 + 7x + C$. You can plug in any real number value for C and the result is a solution to the original differential equation. For example, $y = x^2 + 7x + 5$, $y = x^2 + 7x - \frac{105}{13}$, and $y = x^2 + 7x + 4\pi$ are all possible solutions. These three (plus an infinite number of other equations) make up the family of solutions.

Talk the Talk

Any mathematical solution containing "+C" is called a **family of solutions,** since it gives no one specific answer. It compactly describes an infinite number of solutions, each differing only by a constant. The members of a family of solutions have the same shape, differing only in their vertical position on the coordinate axis. In order to reach a specific solution, the problem will have to provide additional information.

Knowing a family of solutions is sometimes not enough, however. Differential equations are often used as mathematical models to illustrate real-life examples and situations. In such cases, you'll need to be able to find specific solutions to differential equations, but to do so you'll need a little more information up front.

Specific Solutions

In order to determine exactly what C equals for any differential equation solution, you'll need to know at least one coordinate pair of the differential equation's antiderivative. With that information, you can plug in the (x,y) pair and solve for C. To enlighten you on this, I have thrown together a little example for those game show fans out there.

I don't know what it is that makes people (by which I mean my wife and me) so excited to watch other people agonize about winning dishwashers by throwing comically oversized dice, but it doesn't stop us from watching game shows. However, the sudden trend in these programs isn't playing silly games for prizes anymore; instead, they subject the contestants to peril in order to win vast sums of money.

Example 2: A new television game show this fall on FOX TV is making quite a stir. *Terminal Velocity* will suspend contestants by their ankles on a bungee cord. Producers are still working out the details, but one of the show's features will be dropping the participants from the ceiling and allowing them to bounce their way to the studio floor as the length of the bungee is slowly increased. Plus, the audience will throw things at them (like small rocks or maybe piranhas if ratings begin to sag). Suppose that the velocity of a contestant (in ft/sec), for the first 10 seconds of her fall, is given by $\frac{ds}{dt} = -80\sin(2t) - 4$. If the initial position of the doomed individual is 115 feet off the ground, find her position equation.

Solution: We are given a differential equation representing velocity. The solution to the differential equation will then be the antiderivative of velocity, or position. The problem also tells us that the initial position is 115. This means that the contestant's position at time equals 0 is 115, so $s(0) = 115$. We'll use that in a second to find C, but first things first; we need to apply separation of variables to solve the differential equation:

$$\int ds = \int \left(-80\sin(2t) - 4 \right) dt$$
$$s(t) = 40\cos(2t) - 4t + C$$

There you have it—the position equation. Remember that we should get 115 if we plug in 0 for t; make that substitution, and you can find C easily:

$$115 = 40\cos(2 \cdot 0) - 4 \cdot 0 + C$$
$$115 = 40\cos(0) + C$$
$$115 = 40 \cdot 1 + C$$
$$75 = C$$

Therefore, the exact position equation is $s(t) = 40\cos(2t) - 4t + 75$.

You've Got Problems

Problem 2: A particle moves horizontally back and forth across the x-axis according to some position equation $s(t)$; the particle's acceleration (in ft/sec²) is described accurately by the equation $a(t) = 2t + 5 - \sin t$. If you know that the particle has an initial velocity of -2 ft/sec and an initial position of 5 feet, find $v(t)$ (the particle's velocity) and $s(t)$ (the particle's position).

Exponential Growth and Decay

Most people have an intuitive understanding of what it means to have *exponential growth*. Basically, it means that things are increasing in an out-of-control way, like a virus in a horror movie. One infected person spreads the illness to another person, then those two each spread it to another. Two infected people becomes four, four becomes eight, eight becomes sixteen, until it's an epidemic and Jackie Chan has to come in to save the day, possibly with karate kicks.

Talk the Talk

Exponential growth occurs when the rate of change of a population is proportional to the population itself. In other words, the bigger the population, the larger it grows. (With exponential decay, the smaller the population, the more slowly it decreases.) **Logistic growth** begins almost exponentially but eventually grows more slowly and stops, as the population reaches some limiting value.

Truth be told, there are not a lot of natural cases in which exponential growth is exhibited. An exponential growth model assumes that there is an infinite amount of resources from which to draw. In our epidemic example, the rate of increase of the illness cannot go on uninhibited, because eventually, everyone will already be sick. To get around such restrictions, many problems involving exponential growth and decay deal with exciting things like bacterial growth.

Notice that in Figure 20.1 the logistic growth curve changes concavity (from concave up to concave down) about midway through the interval I've drawn. This change in concavity indicates the point at which growth slows.

Figure 20.1

Two kinds of growth. Neither explains that weird mole on your neck.

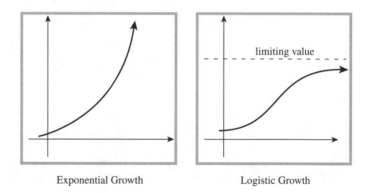

Exponential Growth   Logistic Growth

A more realistic example of growth and decay is *logistic growth*. In this model, growth begins quickly (it basically looks exponential at first) and then slows as it reaches some limiting factor (as our virus could only spread to so many people before everyone was already dead—isn't that a pleasant thought?). Although it is not beyond our abilities do examine logistic growth, it is by far more complicated to understand and model, so we'll stick with exponential growth.

Mathematically, exponential growth is pretty neat. We say that a population exhibits exponential growth if its rate of change is directly proportional to the population itself. Thus, a population P grows (or decays) exponentially if $\frac{dP}{dt}$ and P are in proportion to each other. Without getting into a lot of detail (too late for that, isn't it?), we can say that these two things are in proportion if:

$$\frac{dP}{dt} = k \cdot P$$

Recognize that? It's Example 1! Since we already solved this differential equation earlier in the chapter, we know that a population showing exponential growth has equation:

$$y = Ne^{kt}$$

(I know I used "C" as the constant before, but I like having the "N" there better, because when you read the formula, it looks like the word "naked," and I am immature enough to think that's pretty funny.) In this formula, N represents the beginning or initial popula-

tion, k is a constant of proportionality, and t stands for time. (The e is just Euler's number, which you've undoubtedly seen lurking about in your precalculus work—it's not a variable. Most calculators, even scientific ones, have a button for Euler's number, so you don't have to memorize it.) The y represents the total population after time t has passed. We called that variable P in Example 1, but it means the same thing.

Your first step in the majority of exponential growth and decay problems is to find k, because you will almost never be able to determine what k is based on the problem. Don't even try to guess k—it's rarely, if ever, obvious. For example, if your population increased in this sequence: 2, 4, 8, 16, 32, etc., you may be tempted to think that $k = 2$ since the population constantly doubles, but instead, k ≈ 0.693147.

Example 3: Even after that great movie *Pay It Forward* came out, the movement promoted by the film never really caught on. Its premise was that you should do a big favor for three different people, something they couldn't accomplish on their own. In turn, they would provide favors for three other people, and so on. Instead, a new movement called Punch It Forward is unfortunately catching on. It's the same premise, but with punching instead of favors. On the first day of Punch It Forward, 19 people are involved in the movement. Ten days later, 193

Critical Point _____

You use the same formula for both exponential growth and decay. The only difference in the two is that k will turn out to be negative in decay problems and positive in growth problems.

Critical Point _____

It's easy to see why N represents the initial population. In Example 3, we know that initially (i.e., when $t = 0$), there is a population of 19. Plugging into $y = Ne^{kt}$, that gives us $19 = Ne^{k \cdot 0}$. The exponent for e is 0, and anything (excluding 0) to the 0 power is 1. Therefore, the equation becomes $19 = N \cdot 1$. Instead of doing this in each problem, we just automatically plug the initial value into N.

people are involved. How many people will be involved 30 days after Punch It Forward begins?

Solution: For the sake of ease, we'll assume exponential growth, because the rate of growth of the movement is definitely proportional to the number of members in the movement. Remember, even logistic growth acts like exponential growth in the beginning. Therefore, we'll use the exponential growth and decay formula $y = Ne^{kt}$. N represents the initial population (19). We know that after 10 days have elapsed, the new population is 193. Therefore, when $t = 10$, $y = 193$. Plug all these values in and solve for k:

$$y = Ne^{kt}$$
$$193 = 19e^{10k}$$
$$\frac{193}{19} = e^{10k}$$
$$\ln\left(\frac{193}{19}\right) = 10k$$
$$\frac{\ln\left(\dfrac{193}{19}\right)}{10} = k$$
$$k \approx .231825$$

The exponential growth model is $y = 19e^{.231825t}$. To find out the population after the first 30 days, plug in 30 for t:

$$y = 19e^{.231825(30)}$$
$$y = 19914.2$$

Therefore, approximately 19,914 people have been inducted (and possibly indicted) into the Punch It Forward society. It's a brave new world, my friend.

You've Got Problems

Problem 3: Those big members-only warehouse superstores always sell things in such gigantic quantities. It's unclear what possessed you to buy 15,000 grams of Radon-222 radioactive waste. Perhaps you thought it would complement your 50-gallon barrel of mustard. In any case, it was a bigger mistake to drop it in the parking lot. All radioactive waste has a defined half-life—the period of time it takes for half of the mass of the substance to decay away. The half-life of Radon-222 is 3.82 days. (In other words, 3.82 days after the waste pours out on the asphalt, 7,500 grams remain, and only 3,750 grams 3.82 days after that.) How long will it take for the 15,000 grams of Radon-222 to decay to a harmless 50 grams?

The Least You Need to Know

◆ Differential equations contain derivatives; solutions to basic differential equations are simply the antiderivatives solved for y.

◆ If a problem contains sufficient information, you can find a specific solution for a differential equation; it won't contain a "$+C$" term.

◆ If the growth or decay of a population is proportional to the size of the population, that growth or decay is exponential in nature.

◆ Exponential growth and decay can be modeled with the equation $y = Ne^{kt}$.

Chapter 21

Visualizing Differential Equations

In This Chapter

- ◆ Approximating function values with tangent lines
- ◆ Slope fields: functional fingerprints
- ◆ Using Euler's Method to solve differential equations

Our brief encounter with differential equations is almost at an end. Like ships passing in the night, we will soon go our separate ways, and all you'll have left are the memories. Before you get too nostalgic, though, we have to discuss some slightly more complex differential equation topics.

We'll start with linear approximation, which we actually could have discussed in Chapter 9, because it is basically an in-depth look at tangent lines. However, it is a precursor to a more complex topic called Euler's Method, which is an arithmetic-heavy way to solve differential equations if you can't use separation of variables. It's a good approximation technique to have handy, since there are about 10 gijillion other ways to solve differential equations that we don't know the first thing about at this level of mathematical maturity.

Before we call it quits, we'll also spend some quality time with slope fields. They are exactly what they sound like—a field of teeny little slopes, planted like cabbages. By examining those cabbages, we can tell a lot about the solution to the differential equation. All in all, this chapter focuses on ways to broaden our understanding of differential equations without having to learn a whole lot more mathematics in the process. You have to love that.

Linear Approximation

We have been finding derivatives like mad throughout this entire book. Even though the derivative is the slope of the tangent line, it took us awhile to appreciate why that could be even a remotely useful thing to know. In time, we learned that derivatives describe rates of change and can be used to optimize functions.

Let's add something new to the list about how mind-numbingly useful derivatives are. Take a look at the graph of a function $f(x)$ and its tangent line at the point $(c, f(c))$ in Figure 21.1.

Figure 21.1

The functions may be very close to x = c, but the further away from the point of tangency, the further apart the graphs get. The line, for example, would not give you a good approximate value for the function for x = 0.

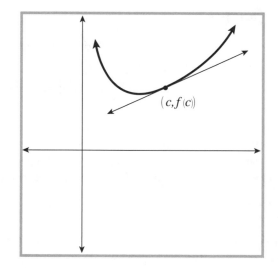

Notice how the graph of f and the tangent line graph get very close to each other in the space right around $x = c$. If you were to plug a value of x very close to c into both functions, you'd almost get the same output. Why is this interesting, though? Well, the derivative of a function is often simpler to deal with than the original function. For example, most polynomials get simpler and simpler with each derivative. Think about it—each time you derive, the degree decreases by one, and if there are any constants, deriving eliminates them.

The whole idea of using a derivative to approximate the original function value may not impress you too much at this point. That's probably because you've done much more impressive things before now. However, we're going to upgrade this whole process in just a few pages when we discuss Euler's Method. You'll be wowed all over again, I promise.

Because the equation of a tangent line to a function has values that usually come very close to the function around the point of tangency, that tangent equation is a good *linear approxi-mation* for the function. No matter how simple a function may be to evaluate, not many functions are simpler than the equation of a line. Further-more, there are some functions that are far from easy to evaluate without a calculator, and linear approximations are very handy in evaluating such functions.

Talk the Talk

A **linear approximation** is the equation of a tangent line to a function that is used to help approximate the function's values lying close to the point of tan-gency.

Example 1: Estimate the value of ln(1.1) using the linear approximation to $f(x) = \ln x$ centered at $x = 1$.

Solution: The problem asks you to center your linear approximation at $x = 1$; this means that you should find the equation of the tangent line to $f(x)$ at that x-value. It's easy to build the equation of a tangent line—you did it way back in Chapter 10. All you need is the slope of the tangent line, which is $f'(1) = \frac{1}{1} = 1$, and the point of tangency, which is $f(1) = \ln 1 = 0$. With this information, use the point-slope equation to build the tangent line:

$$y - y_1 = m(x - x_1)$$
$$y - 0 = 1(x - 1)$$
$$y = x - 1$$

Kelley's Cautions

Remember: A linear approximation is only assumed to give a good estimate for x-values close to the x at which the approximation was centered. In fact, the approximation in Example 1 gives an awful approximation of –.87 for $x = \frac{1}{8}$. The actual value of $\ln\left(\frac{1}{8}\right)$ is –2.079. That's rather inaccurate, even though $\frac{1}{8}$ is pretty close to the center of the approximation!

You might not have any idea what number results when you plug 1.1 into $f(x) = \ln x$, but I bet you could plug 1.1 into the tangent line equation. (I have nothing but faith in you.) You'll get $y = 1.1 - 1 = 0.1$. The actual value of $\ln(1.1)$ is .09531, so our estimation is relatively close.

You've Got Problems

Problem 1: Estimate the value of $\arctan(1.9)$ using a linear approximation centered at $x = 2$.

Slope Fields

Even if you can't solve a differential equation, you can still get a good idea of what the solution's graph looks like. We just learned that a graph's tangent line looks a lot like the graph right around the point of tangency. Well, if we draw little tiny pieces of tangent line all over the coordinate plane, those pieces will show the shape of the solution graph. It's similar to using metallic shavings to determine where magnetic fields lie, or sprinkling fingerprint powder on a table's surface to highlight the shape of the print.

Drawing a *slope field* is a very simple process for basic differential equations, but it can get a bit repetitive. All you have to do is plug points from the coordinate plane into the differential equation. Remember, the differential equation represents the slope of the solution graph, as it is the first derivative. You will then draw a small line segment centered at that point with the slope you calculated.

Talk the Talk

A **slope field** is a tool to help visualize the solution to a differential equation. It is made up of a collection of line segments centered at points whose slope is the value of the differential equation evaluated at those points.

Let's start with a very basic example: $\frac{dy}{dx} = 2x$. We know that the solution to this differential equation is $y = x^2 + C$, a family of parabolas with their vertices on the y-axis. Let's draw the slope field for $\frac{dy}{dx} = 2x$. First, let's pick the fertile field where our slopes will grow and flourish (see Figure 21.2).

At every dot on that field, we are going to draw a tiny little segment. Let's start at the origin. If you plug (0,0) into $\frac{dy}{dx} = 2x$, you get $\frac{dy}{dx} = 2 \cdot 0 = 0$, so the slope of the tangent line there will be 0 (i.e., the line is horizontal). Therefore, draw a small horizontal segment centered at the origin. The substitution was pretty easy—we didn't even have to plug the y-value in, because there was no y in the differential equation.

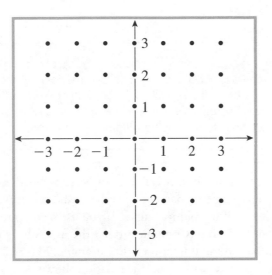

Figure 21.2

Each of the points indicated has coordinates that are integers; this makes the substitution a little quicker and easier.

You can see the job ahead of us—we should do the same thing for every other point we've indicated. Let's do one more together to make sure you've got the hang of this. For fun, I choose the point (1,2)—doesn't that *sound* fun? Plugging that into the differential equation gives you $\frac{dy}{dx} = 2 \cdot 1 = 2$. Therefore, our line segment centered at (1,2) will have a slope of 2.

Go ahead and do the same thing for all the points we indicated; you should end up with something like Figure 21.3.

Critical Point

As a rule of thumb, a slope of 1 means a segment with a 45-degree angle. Greater slopes will be steeper and smaller slopes will be shallower. Negative slopes will fall from left to right, whereas positive slopes rise from left to right.

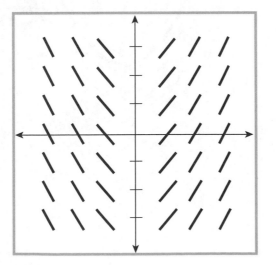

Figure 21.3

A slope field with just a hint of parabola.

> **CAUTION**
>
> **Kelley's Cautions** _____
>
> A slope field will always outline a family of solutions. If you are given a point on the graph of the solution, place your pencil there and follow the paths of the slope segments to get an approximate graph of the specific solution. It's not an exact method of obtaining a graph by any means. Soon, however, you'll learn Euler's Method, and that'll help you find more exact solutions to unsolvable differential equations.

Can you see the shape hiding among all the little twigs? It's not perfect, but these little segments do a pretty good job of outlining the shape of a parabola whose vertex is on the *y*-axis. The slope field traces the shape of its solution curve. If you use a computer to draw the slope field, the shape is even clearer. (Computers don't tire as easily as I do when plugging in points; they don't even mind fractions.) For example, Figure 21.4 is a computer-generated slope field for $\frac{dy}{dx} = 2x$ with a specific solution shown, so you can see how well that the slope field traces the solution.

Figure 21.4

Although one parabola is drawn as a possible solution, it's easy to see that there are a lot of possible parabolas hiding in the woodwork.

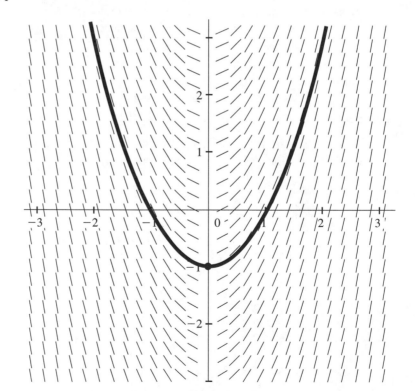

Again, this is a very detailed slope field; the computer calculated slopes at many fractional coordinates as well as integer coordinates. If we know the solution to the differential equation contains the point (0,–1), you get the specific solution, represented by the darkened graph.

Slope fields are most useful when you cannot solve the given differential equation by separation of variables. There are so many differential equations out there that we can't solve at this level of our journey toward mathematical enlightenment, it's good to enlist as many allies as we can.

Critical Point

You can download a fantastic program that draws slope fields on your computer called *Graphmatica* from the kSoft home page, www.graphmatica.com. For more details and other web resources, check out Appendix C.

You've Got Problems

Problem 2: Draw the slope field for $\frac{dy}{dx} = \frac{x+y}{x-y}$ and sketch a possible specific solution to the differential equation containing the point (0,1).

Euler's Method

To understand what *Euler's Method* really accomplishes in the land of differential equations, we need to talk about navigating through the woods. I am not a huge fan of the outdoors, to be perfectly honest. I'm glad to be inside with the air conditioning on, away from flies, ticks, and those ugly little spiders that burrow under your skin and lay eggs in your brain (although this may be a product of watching too many horror movies as a child).

Talk the Talk

Euler's Method is a technique used to approximate values on the solution graph to a differential equation when you can't actually find the specific solution to the differential equation via separation of variables. By the way, Euler is pronounced *OIL-er*, not *YOU-ler*.

This was not the case when I was younger. I always enjoyed tromping around outside and coming in as dirty as possible, covered in mud, sand, grass stains, and mashed bugs. In particular, I enjoyed marching through the woods with my friends. Most of the time, we'd be in areas either my friends or I knew extremely well. We even had crude maps of the woods drawn out, not that we ever actually had to resort to them. In new or unfamiliar woods, however, we'd rely on a compass to direct us to a road we knew: "Okay, if we get lost we'll go west and follow the road home; watch out for those brain-egg spiders along the way."

When we can solve a differential equation using separation of variables, we're given a map to all the correct solutions for that differential equation. In fact, the correct path to follow is analogous to the graph of the equation's solution. Back to our simple example from earlier: If the solution to the differential equation $\frac{dy}{dx} = 2x$ contains the point (3,6), we can easily find the exact solution using separation of variables. The antiderivative will be $y = x^2 + C$, and we can plug in the coordinate pair to find C:

$$6 = 3^2 + C$$
$$6 = 9 + C$$
$$-3 = C$$

Therefore, the exact solution to the differential equation is $y = x^2 - 3$. Now that we have this solution, it is a map that we can use to find other values on the solution graph. For example, it's very easy to determine what the value of $y(4)$ is (i.e., the solution graph's output when you input $x = 4$):

$$y = x^2 - 3$$
$$y(4) = 4^2 - 3 = 16 - 3$$
$$y(4) = 13$$

Once you have the map to the solution graph, it's easy to find the correct y-value corresponding to any x-value. But—and this is a big but—what if you can't solve the differential equation by separation of variables? We use the compass of Euler's Method.

Euler's Method is a way to approximate solutions to a differential equation if you cannot solve it via separation of variables. You'll still be given a point on the solution curve in these problems, but you won't be able to use it to find C. Instead, you'll use it as your reference point ("If you get lost, go west and meet at the road"). From there, you'll take a compass reading ("We should go north—that big tree that looks like Scooby Doo is north from here"). When you have gone a fixed distance, you'll take another compass reading ("Okay, we're at the tree; now we should go northeast to that log with the frog on it"). After every small journey (as you reach each landmark), you'll take a compass reading and make sure that your course is true, and that you're heading in the right direction. After all, you are navigating in unknown lands without a map, with dangerous spiders everywhere. Best to keep checking your compass.

Remember, our big focus in this section has been on this fact: A tangent line has values almost equal to its original function around the point of tangency. This is essential to Euler's Method. Taking compass readings in the woods is analogous to finding the correct derivative for the given function. We'll then step carefully along this slope for a fixed amount of time. If we go too far, the values of the slope will become too different from the values of the function (whose map we don't have). After a bit, we'll make another derivative check and start moving down this new direction. Remember, we don't know where the path is, but by using derivatives, we're staying as close to it as possible.

Before we can actually perform Euler's Method, we need to possess one prerequisite skill. Let's say you are at the point (0,3) and want to walk along a certain line that passes through that point. If that line has slope $m = \frac{2}{5}$, then walking up two units and to the right five units—arriving at the point (2,5)—ensures that we stay on the line. However, what if we only want to go $\frac{1}{3}$ of a unit up? How many units would we go right to make sure we were still on the line?

Example 2: Line n passes through (0,3) and has slope $m = \frac{2}{5}$. Without finding the equation of line n, find the correct y in the coordinate pair $\left(\frac{1}{3}, y\right)$ if that point is also on line n.

Solution: Our new point is exactly $\frac{1}{3}$ to the right of our original point (from 0 to $\frac{1}{3}$), so we can say that the change in x is $\frac{1}{3}$ from the first to

> **Critical Point** _____
>
> In Example 2, you're learning how to use a compass reading $\left(m = \frac{2}{5}\right)$ to walk a short distance $\left(x = \frac{1}{3}\right)$ and yet stay on the correct path.

the second point. Mathematically, this is written $\Delta x = \frac{1}{3}$. All we have to do is find the corresponding Δy to find out how far we should go vertically from our original y-value of 3. Remember that according to the slope equation you learned in your mathematical infancy, slope is equal to the change in y divided by the change in x: $m = \frac{\Delta y}{\Delta x}$. Use this equation to find the correct Δy:

$$m = \frac{\Delta y}{\Delta x}$$

$$\frac{2}{5} = \frac{\Delta y}{\frac{1}{3}}$$

$$\Delta y = \frac{2}{5} \cdot \frac{1}{3} = \frac{2}{15}$$

According to this, we need to go up $\frac{2}{15}$ of a unit from the original y-value of 3. Since $3 + \frac{2}{15} = \frac{45}{15} + \frac{2}{15} = \frac{47}{15}$, the point $\left(\frac{1}{3}, \frac{47}{15}\right)$ is guaranteed to be on the line through (0,3) whose slope is $\frac{2}{5}$.

Now it's time to actually use Euler's Method. Euler's problems give you a differential equation, a starting point, and a value that needs estimating on the solution curve. You'll be told how many steps of what width to use, and you'll take steps of that width using the same method you did in Example 2.

Example 3: Use Euler's Method with three steps of width $\Delta x = \frac{1}{3}$ to approximate $y(3)$ if $\frac{dy}{dx} = x + y$ and the point $(2,1)$ appears on the solution graph.

Solution: It should be clear why the width of the steps is $\Delta x = \frac{1}{3}$—we're stepping from an x-value of 2 to $x = 3$ in three steps. We'll repeat the same process three times, one for each step.

Step One: From $x = 2$ to $x = \frac{7}{3}$ (or $2\frac{1}{3}$)

The correct tangent slope at the point $(2,1)$ is …

$$\frac{dy}{dx} = x + y = 2 + 1 = 3$$

Use this slope to calculate the correct value of Δy:

$$\frac{dy}{dx} = \frac{\Delta y}{\Delta x}$$

$$3 = \frac{\Delta y}{\frac{1}{3}}$$

$$\Delta y = 3 \cdot \frac{1}{3} = 1$$

Kelley's Cautions

The more steps you take (i.e., the smaller the width of each step), the more accurate your final approximation will be. Even with large steps, however, Euler's Method gets messy quickly; the ugly fractions compound, and it's easy to make an arithmetic mistake. It's best to check your work with a calcu-lator.

This tells us that we go up one unit from our original y-value of 1 while stepping right $\frac{1}{3}$ from the original x-value of 2:

$$\left(2 + \frac{1}{3}, 1 + 1\right) = \left(\frac{7}{3}, 2\right)$$

Step Two: From $x = \frac{7}{3}$ to $x = \frac{8}{3}$

Repeat the same process as above, but use a starting point of $\left(\frac{7}{3}, 2\right)$ instead of $(2,1)$. This time, the tangent slope is $\frac{dy}{dx} = x + y = \frac{7}{3} + 2 = \frac{13}{3}$ while Δx remains $\frac{1}{3}$. We should find Δy:

$$\frac{13}{3} = \frac{\Delta y}{\frac{1}{3}}$$

$$\Delta y = \frac{13}{3} \cdot \frac{1}{3} = \frac{13}{9}$$

So, the starting point for the final step will be:

$$\left(\frac{7}{3} + \frac{1}{3}, 2 + \frac{13}{9}\right) = \left(\frac{8}{3}, \frac{31}{9}\right)$$

Step Three: From $x = \frac{8}{3}$ to $x = 3$

This time, the slope of the tangent line is $\frac{dy}{dx} = \frac{8}{3} + \frac{31}{9} = \frac{55}{9}$; again, use it to find that $\Delta y = \frac{55}{27}$. Therefore, we get the point $\left(\frac{8}{3} + \frac{1}{3}, \frac{31}{9} + \frac{55}{27}\right) = \left(3, \frac{148}{27}\right)$. We have our answer. According to Euler's Method, the solution to the differential equation $\frac{dy}{dx} = x + y$ at $x = 3$ is approximately $\frac{148}{27}$, or 5.481.

You've Got Problems

Problem 4: Use Euler's Method with three steps of width $\Delta x = \frac{1}{3}$ to approximate $y(1)$ if $\frac{dy}{dx} = 2x - y$ and if the solution graph passes through the origin.

The Least You Need to Know

◆ A tangent line's values are very close to its original function's values near the point of tangency.

◆ Slope fields are collections of small pieces of tangent lines spread out over the coordinate plane. They trace the graphs of solutions to differential equations.

◆ Euler's Method is used to approximate values for the solutions to differential equations via linear approximation.

Chapter 22

Sequences and Series

In This Chapter

◆ Sequences: more than lists of numbers

◆ Can sequences have limits?

◆ The difference between sequences and series

◆ Understanding very simple series

As they used to say in *Monty Python's Flying Circus*, "Now, for something completely different": It's time for sequences and series. I've always thought it strange that a completely unrelated calculus topic be thrown in at the end of a basic calculus course, but so it has been, and so it shall be. Perhaps it is overstating the matter to say that a brief study in these topics is *completely* unrelated. You'll see some limits (in fact, you'll see limits at infinity, which are always fun), and a smidgen of integration thrown in as well. However, these final chapters are sure to leave a different taste in your mouth than those preceding them.

Sequences and series are often the least understood and most quickly forgotten sections for calculus students. This happens mainly because they fall at the end of the course, when students tire of learning new material, and (sad but true) when teachers are tired of presenting it.

That's not to say that this chapter is difficult to understand, tricky in its methods, or uninteresting by its very nature. In fact, some of the things you'll be able to do are downright fascinating, but it's up to you to stay the course and stay focused. Many a great calculus student has fallen prey to indifference this late in the course. Don't let it happen to you!

What Is a Sequence?

When I was in elementary school, I used to love those little pattern puzzles. You know the ones I'm talking about: "Find the next number in the following sequence: 1, 3, 5, 7, ..." After some careful consideration, you would see that the next number in the pattern would be 36 (just kidding—it's 9). Unbeknownst to you, you were exploring a very basic mathematical *sequence*.

Talk the Talk

A **sequence** is a list of numbers generated by some mathematical rule typically expressed in terms of *n*. In order to construct the sequence, you plug consecutive integer values into *n*.

A sequence is a list or collection of numbers that is generated by some defining rule. It can be written as a list, like our simple sequence above, or in braced notation using the variable *n*. For example, the sequence of odd integers can be generated by the sequence $\{2n-1\}$. See how the sequence develops as I plug in integer values for *n*, beginning with 1:

$$(2(1)-1), (2(2)-1), (2(3)-1), (2(4)-1), (2(5)-1), \ldots$$
$$= 1, 3, 5, 7, 9, \ldots$$

How did I know to use that pattern to generate the odd integers? Well, two times anything must *always* result in an even number, and if you subtract 1 from an even number the result is always odd. Sometimes the trickiest thing about sequences is trying to figure out what the defining rule is, given only a list of the terms in the sequence. All it takes is a little practice (which you'll get in a minute) and a good instinct for patterns.

Kelley's Cautions

You may wonder, "Can I start by plugging in 0 (or any other number besides 1) to get the sequence if it's written in braced notation?" The answer is: Don't sweat the small stuff. When we discuss sequences, we're usually worried about how the one jillionth element behaves, not the first two or three. The main focus is the pattern, not where the pattern starts. This is not so with *series*, however—but more on that at the right time.

Sequence Convergence

Our primary goal when dealing with sequences will be to determine whether or not they *converge*. If a sequence is convergent, the terms in it will approach, but never reach, some limiting real number value. The key word there is "limiting," because we use a limit at infinity to determine whether or not a sequence converges. This makes the process very easy, because you've already dealt with limits at infinity out the wazoo, so there are no new concepts to learn just yet.

Mathematically, we say that the sequence $\{a_n\}$ converges if $\lim_{n\to\infty} a_n$ exists. This makes a lot of sense. In essence, we are looking to see how the sequence acts as n gets very large, which means that we're inspecting the behavior of the sequence far, far down its list of members. If the rule defining the sequence has a limit, then it only makes sense that the sequence created by that rule would also be bounded.

Example 1: Is the sequence $\frac{0}{1}, \frac{1}{4}, \frac{2}{9}, \frac{3}{16}, \frac{4}{25}, \ldots$ convergent or divergent?

Solution: First, we need to come up with some general rule that defines this sequence. The first term ($n = 1$) is $\frac{0}{0}$, while the second term ($n = 2$) is $\frac{1}{4}$. Notice that the numerator is one less than n and the denominator for each term is n^2. The same pattern holds true for the other terms, so our sequence is $\left\{\frac{n-1}{n^2}\right\}$. To see if this converges, we examine $\lim_{n\to\infty} \frac{n-1}{n^2}$. Remember how to do limits at infinity? You can either use L'Hôpital's Rule or compare degrees of the numerator and denominator. Either way, you get a limit of 0. Since the limit exists (i.e., does not increase without bound), the sequence converges.

You've Got Problems

Problem 1: Does the sequence $\left\{\frac{5n^3}{\ln n^2}\right\}$ converge or diverge?

What Is a Series?

A mathematical *series* is very similar to a sequence. However, instead of just listing the numbers as a sequence, you add them together. Series are also a little pickier about where you start and end. To keep from getting confused between sequences and series, I use the World Series as a mnemonic device. How do you get the score of each game of the World Series (or any baseball game, for that matter)? You add the scores together for each individual inning. Thus, series are based on sums, whereas sequences are not.

Talk the Talk

A **series** is the sum of the terms of a sequence. Series are typically written in sigma notation, which indicates which terms of the sequence are the first and last of the series.

Let's take a look at the simple series $\sum_{n=0}^{4} \left(n^2 + 3 \right)$.

Don't get frazzled by the sigma notation (that's what you call that Greek letter out front, for those of you who weren't in fraternities and sororities in college). Those boundaries work a lot like integration boundaries. Start by plugging in the bottom bound (0) and then plug in consecutive integers until you reach the top bound (4), which will be your last term. Remember, this is a series, so you have to add all of the terms together:

$$\sum_{n=0}^{4} \left(n^2 + 3 \right) = \left(0^2 + 3 \right) + \left(1^2 + 3 \right) + \left(2^2 + 3 \right) + \left(3^2 + 3 \right) + \left(4^2 + 3 \right)$$

$$= 3 + 4 + 7 + 12 + 19$$

$$= 45$$

Much of the time in calculus, we don't care about series like $\sum_{n=0}^{4} \left(n^2 + 3 \right)$, because they are finite (they don't contain an infinite number of terms). Finite series are all well and good, but they are akin to scuba diving in a kiddy pool. If you want to see the really exotic fish and big octopi that drag men screaming to their watery graves, we need to examine infinite series—series that have an upper bound of infinity.

Our primary concern with infinite series will be to determine if they converge or diverge, just like sequences. That is to say, if you were to add all the terms in the infinite sequence, would you get an actual answer? It seems bizarre that you can add an infinite number of things together and get, say, $14\frac{2}{3}$, but it happens. Like I told you, we're swimming with the weird and dangerous fish now.

Critical Point

For the rest of this chapter, and all of the next chapter, we're going to be examining infinite series and trying to determine whether or not they converge. In some cases, we'll be able to find the actual sum of the infinite series, but in most, we'll content ourselves with the knowledge that the series does converge, even if we don't know what the eventual sum will be.

Before we slap on the swim fins and the snorkel, we need to discuss one last thing: the *nth term divergence test*. In our bizarre (and strangely persistent) scuba diver metaphor, this test is our speargun. If any infinite series fails this test, it is automatically divergent, and we need not spend one more second on the problem. So, the *n*th term divergence test should be the first thing you apply to any infinite series you see, in the hopes of spearing that problem right in the gut. Once you prove a series divergent, there's nothing else to do other than watch it spiral to the sea floor, on the lookout for more ferocious predators in the water around you.

nth term divergence test: The infinite series $\sum a_n$ is divergent if $\lim_{n \to \infty} a_n \neq 0$.

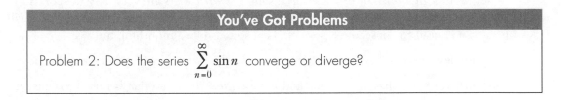

Problem 2: Does the series $\sum_{n=0}^{\infty} \sin n$ converge or diverge?

Think about what this means. If you are adding an infinitely long list of numbers together, there's no way you can get a finite sum unless you're eventually adding 0 (or something close to it) infinitely. For example, if $\lim_{n \to \infty} a_n = \frac{1}{2}$, then you're eventually adding $\frac{1}{2} + \frac{1}{2} + \frac{1}{2} + \frac{1}{2} + \dots$ forever and ever, and that sum will definitely get infinitely large, causing the series to diverge.

Kelley's Cautions

You can *never* use the *n*th term divergence test to prove that a series converges. It can only prove divergence. Consider the series $\sum_{n=1}^{\infty} \frac{1}{n}$ (it is called the "harmonic series"). If you apply the *n*th term divergence test, you see that $\lim_{n \to \infty} \frac{1}{n} = 0$. You *cannot* then conclude that the series converges; in fact, the series *diverges*, in spite of the fact that the sequence $\frac{1}{n}$ *converges*.

Basic Infinite Series

Let me return to my absurd but entertaining metaphor of working with series as scuba diving for a moment. In Chapter 23, you will be plunged into the cold depths of the sea, surrounded by aquatic life of all kinds, and quite possibly solve a very complex crime by noticing sand patterns. (The last of the three events may only happen to you if you are the Scooby Doo gang, Nancy Drew, or one of the Hardy Boys.) Before you can hope to survive in such an unforgiving seascape, you need to log some time in a fishing pond, surrounded by guppies, minnows, and recently deceased teamsters. The series that follow are the guppies of the infinite series world. Let's tangle with them before we bring on the man-eaters.

Geometric Series

Whether in sigma form or expanded as a sum, *geometric series* are very easy to spot. All of the terms will contain a common factor, and once that factor is pulled out, the result will be a value raised to consecutive powers. That sounds like a mouthful, but it's very simple in practice. All geometric series have the form $\sum_{n=0}^{\infty} ar^n$, where a is that constant every term has in common and r is the *ratio*, the value that is raised to consecutive powers to generate the series.

Talk the Talk _____

A **geometric series** has the form $\sum_{n=0}^{\infty} ar^n$, where a and r are constants. When the series is expanded, every term will contain a; consecutive terms will possess consecutive powers of r, which is called the **ratio.** Notice that geometric series begin with $n = 0$. This has the net effect of making the first term a when the series is written as a sum:

$$\sum_{n=0}^{\infty} ar^n = a + ar + ar^2 + ar^3 + \cdots$$

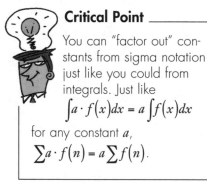

Critical Point _____

You can "factor out" constants from sigma notation just like you could from integrals. Just like

$$\int a \cdot f(x)dx = a \int f(x)dx$$

for any constant a,

$$\sum a \cdot f(n) = a \sum f(n).$$

A geometric series will converge if $0 < |r| < 1$; it will diverge if $|r| \geq 1$. If the series converges, you can even find the sum of the series by plugging into the formula $\frac{a}{1-r}$. Because determining convergence is as easy as looking at the r term, these series couldn't be easier. They're only a little tricky when they are written as a sum rather than in sigma notation, and are only marginally trickier then.

Example 2: Determine whether or not the geometric series $2 + 3 + \frac{9}{2} + \frac{27}{4} + \frac{81}{8} + \dots$ converges. If so, give the sum of the series.

Solution: The problem is very kind in telling us there's a geometric series lurking there. The first term, then, will be the a term, and if you factor it out of every term, you'll get $2\left(1 + \frac{3}{2} + \frac{9}{4} + \frac{27}{8} + \frac{81}{16} + \dots\right)$. Clearly, the ratio in the series is $r = \frac{3}{2}$, because the terms in the expansion are consecutive powers of that fraction. Therefore, we can rewrite the series in true geometric form: $\sum\limits_{n=0}^{\infty} 2\left(\frac{3}{2}\right)^n$. Since $\left|\frac{3}{2}\right| \geq 1$, this series will diverge, and as such, will have no finite sum.

P-Series

If a series has the form $\sum\limits_{n=1}^{\infty} \frac{1}{n^p}$, where p is a constant, it is called a *p-series*. Sometimes *p*-series will try and hide their true identity with additional constants, but any disguise will be transparent. Remember, you can factor constants out of sigma notation. Therefore, even $\sum\limits_{n=1}^{\infty} \frac{7}{3n^{2/3}}$ is a *p*-series. Removing those constants gives you $\frac{7}{3} \cdot \sum\limits_{n=1}^{\infty} \frac{1}{n^{2/3}}$, which is a *p*-series with $p = \frac{2}{3}$.

Determining the convergence of a *p*-series is just as easy a task as it was with geometric series. All you have to do is examine the *p*. (Sounds like something a urologist would do, but don't let that bother you.) A *p*-series will converge if $p > 1$, but will diverge for all other values of *p*. Unlike geometric series, you cannot determine the sum of a convergent *p*-series using a handy formula. In fact, you won't be asked to calculate the sum of an infinite series unless the series is geometric or telescoping (the final simple series that you'll learn before the chapter is through).

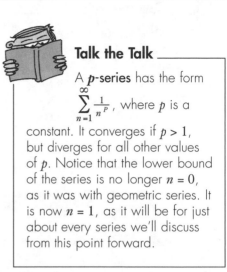

Talk the Talk

A *p*-series has the form $\sum\limits_{n=1}^{\infty} \frac{1}{n^p}$, where *p* is a constant. It converges if $p > 1$, but diverges for all other values of *p*. Notice that the lower bound of the series is no longer $n = 0$, as it was with geometric series. It is now $n = 1$, as it will be for just about every series we'll discuss from this point forward.

Telescoping Series

The key characteristic of *telescoping series* is that all but one or a few terms of the series will cancel out, which makes finding the sum of these series extremely easy. To find the sum, simply write out the series expansion until it's clear what terms disappear, negated by other terms in the series.

Example 3: Find the sum of the convergent telescoping series $\sum_{n=1}^{\infty}\left(\frac{1}{n}-\frac{1}{n+3}\right)$.

Solution: If you expand this series, you get:

$$\left(1-\frac{1}{4}\right)+\left(\frac{1}{2}-\frac{1}{5}\right)+\left(\frac{1}{3}-\frac{1}{6}\right)+\left(\frac{1}{4}-\frac{1}{7}\right)+\left(\frac{1}{5}-\frac{1}{8}\right)+\ldots$$

Talk the Talk

A **telescoping series** contains an infinite number of terms and their opposites, resulting in almost all of the terms in the series canceling out. Telescoping series, by their very nature, must contain a subtraction sign.

If we regroup the terms, you get:

$$1+\frac{1}{2}+\frac{1}{3}+\left(\frac{1}{4}-\frac{1}{4}\right)+\left(\frac{1}{5}-\frac{1}{5}\right)+\ldots$$

Every term greater than $\frac{1}{3}$ will be canceled out by its opposite somewhere along the lifespan of the series. Therefore, the sum of the series is simply the sum of the three terms that remain:

$$1+\frac{1}{2}+\frac{1}{3}=\frac{11}{6}$$

You've Got Problems

Problem 5: Calculate the sum of the series $\sum_{n=1}^{\infty}\left(\frac{1}{n}-\frac{1}{n+1}\right)$.

The Least You Need to Know

◆ A sequence is a list of numbers based on some defining rule. A sequence converges if the limit at infinity of its defining rule exists.

◆ A series is the sum of a specific number of terms of a sequence, as defined by lower and upper boundaries. If the upper boundary is infinity, the given series is said to be infinite.

◆ Geometric series have the form $\sum_{n=0}^{\infty}ar^{n}$ and converge only if $0<|r|<1$; the sum of a convergent geometric series is $\frac{a}{1-r}$.

◆ *P*-series have the form $\sum_{n=1}^{\infty}\frac{1}{n^{p}}$ and converge only if $p>1$.

Chapter 23

Infinite Series Convergence Tests

In This Chapter

- ◆ A link to integration?
- ◆ Comparing series: "Why can't you be like your sister?"
- ◆ Disarming series using the Ratio and Root Tests
- ◆ Investigating alternating series

When I first taught series as a high school teacher, I hated them. I loathed them. I don't now, but back then, it was a different story. In fact, when the thought of infinite series crossed my mind, my face twisted into an involuntary sneer. It could, however, have something to do with the circumstances. I was the fill-in guy for the *real* calculus teacher, because she was pregnant.

I didn't remember a single derivative, integral, or limit from calculus, and now I had to teach infinite series to a bunch of students who'd probably rather have seen me disemboweled than listen to explanations for concepts I only half understood myself. But you probably already know what hating an occasional calculus topic feels like, and I'm preaching to the choir. The good news is that you have a strong and secure calculus background behind you, so series convergence tests will be very easy to understand.

Which Test Do I Use?

As you may have guessed, this chapter is full of tests used to determine whether or not infinite series converge. We won't be able to find sums of any infinite series anymore—these series are too complex for that—so we'll settle for simply knowing whether or not that mystery sum exists.

Like integration problems, infinite series require numerous techniques. Sometimes, you'll be able to tell what convergence test to use just by looking at a series, but much of the time, you'll have to do a little experimenting to find one that works. Have you ever noticed that it's easy to choose between things if only given a few options? Well, this chapter adds six additional convergence tests to the three from last chapter, and it can be tricky to decide which to choose. Just like everything else in calculus, however, practice makes perfect. Eventually, you'll develop an instinct that nudges you toward the correct method to use based on the look of the problem.

The Integral Test

In Chapter 18, you learned how to calculate improper integrals by using limits. One of the major causes of an improper integral was an infinite boundary. Can you see any relation to infinite limits? Look at the two problems that follow:

$$\int_1^\infty \frac{1}{1+2n}\,dn \qquad \sum_{n=1}^\infty \frac{1}{1+2n}$$

Talk the Talk

The **Integral Test** states that the positive series $\sum_{n=1}^\infty a_n$ converges if the improper integral $\int_1^\infty a_n\,dn$ has a finite value. The opposite is also true—if the definite integral is infinitely large, the series diverges. It doesn't make a lot of sense to apply the Integral Test to a function that is increasing along its domain, because the area trapped beneath that curve will not be finite.

These problems look almost identical, just like Superman and mild-mannered newspaper reporter Clark Kent. In fact, the series on the right will converge if the integral on the left converges (i.e., has a numeric value). However, if the integral on the left increases without bound, then so does the infinite sum on the right. This correlation between the convergence of an infinite series and its corresponding improper integral is called the *Integral Test*.

You can only apply the Integral Test to series that exclusively contain positive terms (called positive series). We'll be able to deal with series containing negative and alternating positive and negative terms with other tests.

Example 1: Use the Integral Test to determine the convergence of $\sum\limits_{n=1}^{\infty}\dfrac{1}{1+2n}$.

Solution: As I've already mentioned, this series will converge if $\int_{1}^{\infty}\dfrac{1}{1+2n}dn$ is a finite number. To evaluate the definite integral, you need to replace the infinite boundary with a limit as we did in Chapter 18:

$$\int_{1}^{\infty}\frac{1}{1+2n}\,dn = \lim_{a\to\infty}\left(\int_{1}^{a}\frac{1}{1+2n}\,dn\right)$$

You can integrate this fraction via u-substitution with $u = 1 + 2n$:

$$\lim_{a\to\infty}\left(\frac{1}{2}\int_{3}^{1+2a}\frac{1}{u}\,du\right)$$

$$=\lim_{a\to\infty}\left(\frac{1}{2}\left(\ln|u|\right)\Big|_{3}^{1+2a}\right)$$

$$=\lim_{a\to\infty}\left(\frac{1}{2}\ln|1+2a|-\frac{1}{2}\ln 3\right)$$

As a gets infinitely large, so does $\ln|1+2a|$, so the integral increases without bound and is divergent. Therefore, the series $\sum\limits_{n=1}^{\infty}\dfrac{1}{1+2n}$ is also divergent.

> **CAUTION**
>
> **Kelley's Cautions**
>
> If $\int_{1}^{\infty}a_{n}\,dn = B$, where B is a constant, you can assume that $\sum\limits_{n=1}^{\infty}a_{n}$ converges, but you *cannot* assume that the sum of the series is B.

You've Got Problems

Problem 1: Use the Integral Test to determine whether or not $\sum\limits_{n=1}^{\infty}\dfrac{\ln n}{n}$ converges.

The Comparison Test

Though you may not have realized it, you are already the master of a number of very different-looking series. However, in the fabulous world of math, as we both know, things very rarely work out easily. Geometric series were very easy to work with, but how often are you going to see something in the specific form $\sum_{n=0}^{\infty} 3 \cdot \left(\frac{1}{4}\right)^{n}$? More often than not, you'll be given series that are *almost* geometric, but not quite. What a tease! We have a great technique that is proven to work with geometric series and along comes this series that smells geometric, feels geometric, but doesn't act geometric.

Critical Point

As you can probably tell from the sibling metaphor, the Comparison Test works best when a series resembles a known series type, but is off by just a bit. Notice that both the original series and the series to which you compare it must be positive.

If you have siblings, you know what it's like to be compared to them. "Why can't you get good grades like your sister?" and "I wish you could rebuild a transmission with the clarity and sense of purpose that your brother Hank can" probably sound familiar to you. Don't look now—you've turned into your parents: "Why can't you be a geometric series like that one?"

Fortunately, comparing series is not so spirit-crushing as comparing people to one another. In fact, you can use the merits and strengths of the "good" series to show that the well-meaning but a little misguided sibling series converges or diverges also.

The Comparison Test (also called the Direct Comparison Test): Given two positive infinite series $\sum a_n$ and $\sum b_n$, such that every term of $\sum a_n$ is less than or equal to the corresponding term in $\sum b_n$,

1) If $\sum b_n$ converges, then $\sum a_n$ converges.

2) If $\sum a_n$ diverges, then $\sum b_n$ diverges.

Think about it this way: If some series, called *B*, eventually converges, and every term in a series *A* is smaller than the matching term in *B* (e.g., the fifth term in *A* is smaller than the fifth term in *B*), then *A must* also converge. Smaller numbers must add up to a smaller sum. The divergence clause of the Comparison Test works the same way. A series larger (on a term-by-term basis) than a divergent series must also be divergent.

Example 2: Use the Comparison Test to show that $\sum_{n=0}^{\infty} \frac{4^n+2}{3^n}$ is divergent.

Solution: Forget about the "+2" for now. If it weren't for that dang 2, you'd have the series $\sum_{n=0}^{\infty} \left(\frac{4}{3}\right)^n$, which is a geometric series with $a = 1$ and $r = \frac{4}{3}$. Notice that our slightly mutated series containing the "+2" term will be larger than its corresponding geometric series. The denominators in both series are equal, but the numerator in the mutated series will be two larger than its counterpart for every term in the series:

$$\sum_{n=0}^{\infty} \frac{4^n + 2}{3^n} = \frac{6}{3} + \frac{18}{9} + \frac{66}{27} + \cdots$$

$$\sum_{n=0}^{\infty} \frac{4^n}{3^n} = \frac{4}{3} + \frac{16}{9} + \frac{64}{27} + \cdots$$

So, $\sum_{n=0}^{\infty} \frac{4^n + 2}{3^n}$ is larger than the divergent geometric series $\sum_{n=0}^{\infty} \left(\frac{4}{3}\right)^n$. (The latter is divergent because $\left|\frac{4}{3}\right| \geq 1$.) Therefore, $\sum_{n=0}^{\infty} \frac{4^n + 2}{3^n}$ is also a divergent series according to the Comparison Test.

You've Got Problems

Problem 2: Use the Comparison Test to determine the convergence of $\sum_{n=1}^{\infty} \frac{1}{n^4 + 3}$.

The Limit Comparison Test

Fractions with n's being raised to exponential powers in both the numerator and denominator make good candidates for this Limit Comparison Test. Just like in the Comparison Test, you'll have to design a comparison series. Let's say we're given the sloppy series $\sum a_n$. The comparison series you'll create will ignore all the terms in the numerator and denominator of $\sum a_n$ except for the n term to the highest degree. Let's call the comparison series we created $\sum b_n$. Now, divide $\sum a_n$ by $\sum b_n$ and evaluate the limit as n approaches infinity. If the limit exists (i.e., is a positive, finite number), then both series act the same way—that is to say, both series either converge or diverge. Now that you've got the general idea, here's what it looks like mathematically.

Kelley's Cautions

The Limit Comparison Test only works for positive series, just like the Integral Test and the Comparison Test. Also, note that the limit in the Limit Comparison Test, if it exists, only tells you whether both series converge or diverge. The limit is not equal to the sum of the series, just like the value of the definite integral in the Integral Test was not equal to the sum of its corresponding series.

The Limit Comparison Test: Given the positive infinite series $\sum a_n$ and $\sum b_n$, if:

$$\lim_{n \to \infty} \frac{a_n}{b_n} = N$$

where N is a positive and finite number, then $\sum a_n$ and $\sum b_n$ either both converge or both diverge.

Example 3: Use the Limit Comparison Test to determine the convergence of $\sum_{n=1}^{\infty} \frac{4n+1}{n^2+3n-2}$.

Solution: This fraction contains n being raised to various powers in both parts of the fraction, so it's a good candidate for the Limit Comparison Test. To generate its comparison series, take only the highest powers of n in the numerator and denominator, and ignore the rest (even the coefficients of those n terms). You get a comparison series of $\sum_{n=1}^{\infty} \frac{n}{n^2}$ or $\sum_{n=1}^{\infty} \frac{1}{n}$. Now, let's evaluate the limit at infinity of the terms of the original series divided by the terms of the new series:

$$\lim_{n \to \infty} \frac{\dfrac{4n+1}{n^2+3n-2}}{\dfrac{1}{n}}$$

Dividing a fraction by another fraction is equivalent to multiplying the top fraction by the reciprocal of the bottom one:

$$\lim_{n \to \infty} \frac{4n+1}{n^2+3n-2} \cdot \frac{n}{1}$$

$$\lim_{n \to \infty} \frac{4n^2+n}{n^2+3n-2} = 4$$

Because the limit is a positive finite number, both series converge or diverge. Clearly, the comparison series $\sum_{n=1}^{\infty} \frac{1}{n}$ is a divergent p-series. Since the comparison series diverges, both series diverge.

You've Got Problems

Problem 3: Determine the convergence of $\sum_{n=1}^{\infty} \frac{\sqrt[3]{n}}{n^2+1}$ using the Limit Comparison Test.

The Ratio Test

The Ratio Test is very useful for series whose terms get really big really fast. If the series in question has exponents, or (even better) n exponents or factorials, the Ratio Test is one-stop shopping for all your convergence needs. Both this and the Root Test (coming up next, so don't change that dial!) work like a Magic 8 Ball—ask it whether or not the series converges, give it a good shake, and see what happens. Will you understand how to do this? "Signs point to yes!"

The Ratio Test: If $\sum a_n$ is an infinite series of positive terms, and $\lim_{n \to \infty} \frac{a_n + 1}{a_n} = L$, then:

1) $\sum a_n$ converges if $L < 1$,

2) $\sum a_n$ diverges if $L > 1$ or if $L = \infty$, and

3) If $L = 1$, the Ratio Test can draw no conclusion. The series may converge, may diverge, or may just be waiting for the perfect moment to punch you in the nose and scuttle back into the shadows. You'll have to use another technique to test for convergence; this one came up snake eyes.

Critical Point

Series containing factorials are prime candidates for the Ratio Test. Remember what a factorial is? It's a little exclamation point next to a number, like this: 5!. Mathematically, $5! = 5 \cdot 4 \cdot 3 \cdot 2 \cdot 1 = 120$. A factorial is basically the product of the number and every integer that comes before it, down to and including 1.

Critical Point

The Ratio Test contains the term a_{n+1}, which is just the series formula with $(n + 1)$ plugged in for n. In essence, we're dividing a term of the series into the next consecutive term and examining the result.

Example 4: Use the Ratio Test to determine the convergence of $\sum\limits_{n=1}^{\infty} \frac{n!}{2^n}$.

Solution: As n approaches infinity, both the numerator and denominator will get large quickly. So, we'll use the Ratio Test. Plug $(n + 1)$ into n and then multiply the result by the reciprocal of the general term (it's the same as dividing by the general term):

$$\lim_{n \to \infty} \frac{(n + 1)!}{2^{n+1}} \cdot \frac{2^n}{n!}$$

Watch carefully. I am going to do some tricky rewriting here. My goal is to split up the factorial in the numerator and the big-exponented 2 in the denominator:

$$\lim_{n \to \infty} \frac{(n+1)(n!)2^n}{(n!)2^n \cdot 2^1}$$

$$= \lim_{n \to \infty} \frac{n+1}{2}$$

This fraction will increase without bound as n approaches infinity; according to the Ratio Test, this indicates divergence of the series.

Here are the justifications to my rewriting in Example 4. First of all, $(n + 1)! = (n + 1) \cdot n!$ for the same reason that $5! = (5) \cdot 4! = (5) \cdot 4 \cdot 3 \cdot 2 \cdot 1$. Second, $2^{n+1} = 2^n \cdot 2^1$ because we know that $x^{a+b} = x^a \cdot x^b$.

You've Got Problems

Problem 4: Use the Ratio Test to determine the convergence of $\sum_{n=1}^{\infty} \frac{n \cdot 3^n}{n!}$.

The Root Test

The Root Test is the sister of the Ratio Test, because it also examines a limit at infinity and applies the *exact same three conditional results* based on how the limit compares to 1. However, instead of examining the limit of a ratio, it examines the limit of the nth root. Therefore, this test is best used when all the pieces of the series at hand are raised to the nth power. Hallelujah! Finally, a series convergence test with an obvious niche! Remember, just look for everything to the nth power, and you're on Easy Street.

The Root Test: If $\sum a_n$ is an infinite series of positive terms, and $\lim_{n \to \infty} \sqrt[n]{a_n} = L$, then

1) $\sum a_n$ converges if $L < 1$,

2) $\sum a_n$ diverges if $L > 1$ or if $L = \infty$, and

3) If $L = 1$, the Ratio Test can draw no conclusion (just like the Ratio Test)

Example 5: Use the Root Test to determine the convergence of $\sum_{n=1}^{\infty} \left(\frac{n^2 + 2n - 1}{5n^2 + 16n - 12} \right)^n$.

Solution: The entire enchilada is raised to the nth power, which indicates that the Root Test is the way to go:

$$\lim_{n \to \infty} \sqrt[n]{\left(\frac{n^2 + 2n - 1}{5n^2 + 16n - 12} \right)^n}$$

Notice how the nth root and the nth power cancel one another out, so all you're left with is a very simple limit at infinity:

$$\lim_{n \to \infty} \frac{n^2 + 2n - 1}{5n^2 + 16n - 12} = \frac{1}{5}$$

Because $\frac{1}{5} < 1$, the series converges.

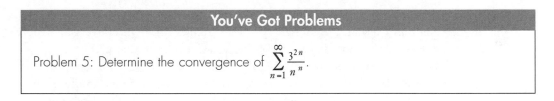

You've Got Problems

Problem 5: Determine the convergence of $\sum_{n=1}^{\infty} \frac{3^{2n}}{n^n}$.

Series with Negative Terms

Every convergence test we've looked at so far has stipulated that it only works when all of the terms of the series are positive. Well, let me tell you something: It's hard to be positive all the time. Occasionally, you just need to be negative. On those mornings I can't find my car keys and stub my toe on the coffee table while looking for them, running around like a lunatic because I'm already late, I understand the need to deal with a series containing negative items. Before we close out this chapter, we'll learn two coping mechanisms for dealing with just such a series.

The Alternating Series Test

Most of the series you'll encounter that contain negative terms are *alternating series*, which are series whose consecutive terms alternate between positive and negative. In other words, every other term has the same sign. Given an alternating series, you can tell whether or not it converges by applying a relatively simple, two-part test.

The Alternating Series Test: If $\sum a_n$ is an alternating series and the following two conditions are met, then $\sum a_n$ converges:

1) Every term of the series is less than or equal to the term preceding it.

2) $\lim_{n \to \infty} a_n = 0$

Talk the Talk

Alternating series are series whose consecutive terms alternate between positive and negative. For example,

$$\sum_{n=1}^{\infty} \frac{(-1)^n}{n} = -1 + \frac{1}{2} - \frac{1}{3} + \frac{1}{4} - \frac{1}{5} + \cdots$$

is an alternating series. Usually, alternating series contain $(-1)^n$ or $(-1)^{n+1}$. It is this piece that causes the signs of the series to alternate, since -1 to an odd power is negative but is positive when raised to an even power.

So, if you've got an alternating series and you want to determine its convergence, here's what you do. First, write out a few terms of the series. Is it clear that every term is smaller than or equal to the one before it (ignoring the positive and negative signs)? If so, that's great—we can move on to the next step. (If not, then you'll have to try absolute convergence, which we'll talk about later.) Second, you have to make sure the series (again ignoring the signs) has a limit at infinity of 0.

Example 6: Determine the convergence of $\sum_{n=2}^{\infty} \frac{(-1)^{n+1}}{n^2 - 1}$.

Solution: Start by writing out a few terms of the series:

$$\sum_{n=2}^{\infty} \frac{(-1)^{n+1}}{n^2 - 1} = -\frac{1}{3} + \frac{1}{8} - \frac{1}{15} + \frac{1}{24} - \frac{1}{35} + \cdots$$

Man, oh man, that denominator is getting big quickly. This is causing each term of the series to get smaller and smaller, so the first condition of the Alternating Series Test is met. (Make sure to ignore the positive and negative signs when you check for term shrinkage.) Now, one more obstacle remains. We need to evaluate the limit (as n approaches infinity) of the series formula without the $(-1)^{n+1}$ piece. Just like in the first part of the Alternating Series Test, the presence of negative signs is irrelevant:

$$\lim_{n \to \infty} \frac{1}{n^2 - 1} = 0$$

As long as that limit equals 0 (which it does), it satisfies the second condition of the Alternating Series Test. Now that both conditions are satisfied, we can conclude that $\sum_{n=2}^{\infty} \frac{(-1)^{n+1}}{n^2 - 1}$ converges.

Absolute Convergence

Sometimes, the Alternating Series Test fails you. In such cases, you'll be forced to examine the *absolute convergence* of the series. Basically, you'll ignore the signs of the series and use one of the tests we've learned already in this chapter. Ignoring the signs (i.e., assuming that all the terms are positive) is key, since the rest of our convergence tests only worked on positive series.

Let's say you start with a series $\sum a_n$, which contains some negative terms. Note that this doesn't have to be an alternating series—it just has to contain at least one negative term. If the series $\left(\sum |a_n|\right)$—which is the original series with all minus signs changed to plus signs—converges, then we say that $\sum a_n$ converges *absolutely*, meaning that regular old $\sum a_n$ is automatically a convergent series, even though you couldn't reach that conclusion via the Alternating Series Test. If, however, $\sum |a_n|$ does not converge

Talk the Talk

A series $\sum a_n$ exhibits **absolute convergence** if $\sum |a_n|$ converges. It is a method of determining whether or not a series containing negative terms converges if you cannot use the Alternating Series Test.

using one of our original tests, then we have no idea whether or not $\sum a_n$ converges, and we all go home dejected, hoping to find something good to watch on TV.

Example 7: Determine the convergence of $\displaystyle\sum_{n=1}^{\infty} \frac{(-1)^{n+1} \cdot n^2}{n!}$.

Solution: This is definitely an alternating series, and the terms definitely shrink as n grows, but $\displaystyle\lim_{n\to\infty} \frac{n^2}{n!} = \frac{\infty}{\infty}$.

Odd—we want that limit to equal 0, but it comes out indeterminate. Since the Alternating Series Test sort of fizzles on us (we can't tell if the limit is 0 or not), we'll see if the series converges absolutely. In other words, we'll look at the series:

$$\sum_{n=1}^{\infty} \frac{n^2}{n!}$$

Both pieces of the fraction increase quickly; therefore, we'll use the Ratio Test to see if the series converges:

$$\lim_{n \to \infty} \frac{(n+1)^2}{(n+1)!} \cdot \frac{n!}{n^2}$$

$$= \lim_{n \to \infty} \frac{n^2 + 2n + 1}{(n+1) \cdot n^2}$$

$$= \lim_{n \to \infty} \frac{n^2 + 2n + 1}{n^3 + n^2} = 0$$

Since $0 < 1$, $\displaystyle\sum_{n=1}^{\infty} \frac{n^2}{n!}$ converges. Therefore, $\displaystyle\sum_{n=1}^{\infty} \frac{(-1)^{n+1} \cdot n^2}{n!}$ automatically converges also.

You've Got Problems

Problem 7: Determine the convergence of $\displaystyle\sum_{n=0}^{\infty} \frac{(-1)^n \cdot 4 \cdot 2^n}{3^n}$.

The Least You Need to Know

◆ You can determine the convergence of a series when the general term is easily integrated using the Integral Test.

◆ If your series closely resembles a far simpler series, you can apply the Comparison or Limit Comparison Test to determine whether or not the series converges.

◆ The Ratio and Root Tests require that you calculate a limit at infinity. How this limit compares to the number 1 tells you whether or not the given series converges.

◆ If your series contains alternating negative and positive terms, you have to apply the Alternating Series Test to see if it converges; if that doesn't work, test to see whether the series converges absolutely.

Chapter 24

Special Series

In This Chapter

◆ Can series be functions?

◆ Power, Taylor, and Maclaurin series

◆ Finding the radius and intervals of convergence

◆ Building approximation polynomials

With Chapter 23 under your belt, you know a thing or two about series. Basically, you are a superhero, able to determine series convergence in a single bound, faster than a speeding limit, more powerful than the Power Rule. In this chapter, you'll learn a few more superhero skills, and then you're off to battle villains, desperados, and evil geniuses that build diabolical weather machines in order to further their plans of world domination.

Your final skills are based on your knowledge of series. There's a new twist, however. We're going to start using series that contain variables, so that the result is not merely a string of numbers, but is function instead. Some series can be used to approximate function values, like we did with linear approximation a little while back. However, these power series do a much better job of approximating functions, and in fact can sometimes get the exact function value, even if the function is a little complicated.

The grand finale will be a brief study in approximation polynomials that result from things called Taylor and Maclaurin series. These are also used to approximate function values, are based on a series definition, and are much better than linear approximations for estimating function values. Practically speaking, there's a little bit more memorizing ahead of you, but none of the math is new or complex. The real question is this: Is this the first hour of the last day in your mathematical journey, or merely the last hour of the first day? There is still much to learn ….

Power Series

We have done a lot of approximating in this book. Many of our techniques began with a rough way to do things, and then we refined it into a snazzy and more accurate methodology. For example, before we jumped into finding exact area beneath curves using the Fundamental Theorem of Calculus, we plodded along, approximating area by shoving known shapes underneath the curves whose areas we knew—things like rectangles, trapezoids, and (to break up the monotony) howler monkeys.

During our approximation techniques, it became clear that the more calculations involved in the process, the more accurate the prediction would be. For example, more rectangles beneath a curve meant a more accurate Riemann sum. Smaller Δx steps along a differential equation during Euler's Method meant a better solution to that differential equation.

In fact, the best way to get a good approximation is to use an infinite number of steps. Thanks to infinite series, we now possess a tool with such capabilities; our approximations can get ridiculously close. *Power series* are infinite series containing x's centered around some value c that give insanely close function approximations for x-values near c, sometimes even producing the exact function value, even if the function is not a simple one. They have the form $\sum_{n=0}^{\infty} a_n(x-c)^n$, where a_n is some formula containing n's, representing the coefficient of each term.

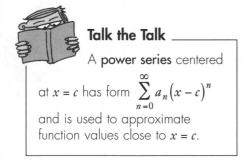

Talk the Talk

A **power series** centered at $x = c$ has form $\sum_{n=0}^{\infty} a_n(x-c)^n$ and is used to approximate function values close to $x = c$.

Radius of Convergence

Like it or not, your only focus will be to determine where power series converge. The good news is they always converge *somewhere*, but the bad news is that there are three "somewheres" where the power series could converge:

- ◆ Only at the value *c* where the series is centered
- ◆ At all real numbers within some radius *r* (called the *radius of convergence*) of the center of the series
- ◆ Everywhere

Critical Point

It's all well and good that power series can theoretically approximate function values. However, that's pretty advanced stuff. Let's be frank about what *you'll* actually be doing. You won't be designing power series. You probably won't even know what functions, if any, the power series are trying to approximate. You're only concerned with determining where power series converge. That's it. Just like last chapter. So, even though the grandiose scheme of things sounds impressive with power series, your role is pretty mundane. However, "mundane" usually means "not too hard" in math, so it's really a blessing in disguise.

Talk the Talk

If a power series is centered around $x = 5$ and has a radius of convergence of 3, then that power series converges on the interval $(5 - 3, 5 + 3) = (2,8)$. We don't know if the series converges at the endpoints of the interval yet—we'll talk about that in the next section. We can describe the same interval mathematically like this: $|x - 5| < 3$. In other words, a power series centered at $x = c$ with radius of convergence r converges for all x that satisfy $|x - c| < r$. In simpler terms, if the **radius of convergence** is r, then the power series will converge for any x between $c - r$ and $c + r$.

In Example 1, we'll work through a certain power series to find the correct radius of convergence. Here are a few important things I want you to watch for as the problem progresses:

- ◆ Using the Ratio Test for absolute convergence with power series
- ◆ Finding the radius of convergence by forcing your result into the form $|x - c| < r$

Critical Point

The radius of convergence for a power series that converges only at its center is $r = 0$. If the power series converges for all real numbers, the radius is infinite. In Example 1, we'll get the remaining case of convergence: a real number radius.

Example 1: Find the radius of convergence for the power series $\sum\limits_{n=0}^{\infty} \dfrac{n^2(x-1)^n}{4^n}$.

Solution: Remember, a power series has form $\sum\limits_{n=0}^{\infty} a_n(x-c)^n$, so we have a power series where $a_n = \dfrac{n^2}{4^n}$ and $c = 1$. Therefore, this power series can approximate values for some function at least for the value of $x = 1$, and perhaps for some values close by. How close must those x-values be to make this series converge? That's the question of the day. Since power series (by definition) contain things to the n power, you use the Ratio Test to determine where they converge. In addition, you will always examine absolute convergence, i.e., you take the absolute value of the series. So even if this were an alternating series, or just a series with negative terms, you'd follow the same process. Now, on to the Ratio Test, just as we did it last chapter:

$$\lim_{n\to\infty} \left| \frac{(n+1)^2(x-1)^{n+1}}{4^{n+1}} \cdot \frac{4^n}{n^2(x-1)^n} \right|$$

$$= \lim_{n\to\infty} \left| \frac{(n^2+2n+1)(x-1)}{4n^2} \right|$$

Notice that $\dfrac{n^2+2n+1}{4n^2}$ approaches $\dfrac{1}{4}$ as n approaches infinity. The $(x-1)$ term is unaffected by n's infinite growth, since it contains no n's. Therefore, the limit equals $\left|\dfrac{1}{4}(x-1)\right|$. So, what now? Well, the Ratio Test says that this limit must be less than 1 in order for the series to converge, so for convergence we need:

$$\left| \frac{1}{4}(x-1) \right| < 1$$

At this point, we almost have the form $|x-c| < r$ (remember that c is 1). In order to achieve that form, multiply both sides by 4, and you get $|x-1| < 4$. This tells you that the radius of convergence is 4. The series $\sum\limits_{n=0}^{\infty} \dfrac{n^2(x-1)^n}{4^n}$ will thus converge on the interval $(c-r, c+r) = (1-4, 1+4) = (-3, 5)$. This means if you substitute any x on that interval into the x spot in the series, the result will be a convergent series.

You've Got Problems

Problem 1: Find the radius of convergence for the power series . $\displaystyle\sum_{n=0}^{\infty} \frac{5^n x^n}{n!}$

Interval of Convergence

In Example 1, we determined that the series $\displaystyle\sum_{n=0}^{\infty} \frac{n^2(x-1)^n}{4^n}$ converges inside the interval $(-3,5)$. However, the series might also converge at the endpoints of the interval, $x = -3$ and $x = 5$. By testing the endpoints, you can determine the *interval of convergence*.

Example 2: On what interval does the series $\displaystyle\sum_{n=0}^{\infty} \frac{n^2(x-1)^n}{4^n}$ (from Example 1) converge?

Solution: We already know the radius of convergence is 4 and that the series converges inside the interval $(-3,5)$. All that's left is to determine the convergence at the endpoints.

Talk the Talk

The **interval of convergence** for a power series centered at c is found once you have determined the radius of convergence, r. You have to plug the endpoints of the interval of convergence ($c - r$ and $c + r$) into the series for x separately to see if the resulting series converges or diverges. The series may converge for both, for neither endpoint, or maybe just for one of them.

Step One: Test $x = -3$

Plug $x = -3$ into the series and you get:

$$\sum_{n=0}^{\infty} \frac{n^2(-3-1)^n}{4^n} = \sum_{n=0}^{\infty} \frac{n^2(-4)^n}{4^n}$$

Now you can rewrite the series to get $\displaystyle\sum_{n=0}^{\infty} n^2\left(\frac{-4}{4}\right)^n = \sum_{n=0}^{\infty} n^2(-1)^n$. The series diverges according to the nth term divergence test. Our conclusion: The series diverges when $x = -3$.

Step Two: Test $x = 5$

When you plug $x = 5$ into the series and simplify, you get $\displaystyle\sum_{n=0}^{\infty} n^2(1)^n$, which also diverges according to the nth term divergence test. Since the series diverged at both of the endpoints we tested, our final conclusion is that $\displaystyle\sum_{n=0}^{\infty} \frac{n^2(x-1)^n}{4^n}$ has an interval of convergence of $(-3,5)$; neither endpoint is included in the interval.

Maclaurin Series

I love fast food. It's one of my true weaknesses. There's something really great about asking for a cheeseburger, and having one handed to you within 45 seconds. True, it's a little off-putting when that cheeseburger has emitted so much grease by the time you reach your car that the bag carrying your food is almost transparent. However, that cruel mistress who is convenient food sings her siren song, and I keep wandering back.

When you go to a fast-food joint, you don't really expect to get home-cooked food. (No one at home wraps their food in paper, puts side items in cardboard envelopes, or keeps food warm under high-powered red heat lamps.) Nothing ever quite tastes like home-made. This is both good and bad. I've always thought that fast-food French fries and soft drinks taste infinitely better than their homebound counterparts, whereas the hamburgers are always a lot better right off your backyard grill.

Even if the taste is not the same, however, fast food is a relatively good approximation of home-cooked food. Furthermore, fast food takes little to no preparation, and you can experience a whole spectrum of different fast foods from different restaurants without having to learn additional skills. You order the same way at Taco Bell, McDonald's, and Wendy's—as long as you have money, you are only seconds away from a tasty and grease-laden entreé.

Critical Point

Next you'll learn how to use Taylor series, which works almost exactly like Maclaurin series, except that you can center them at any x-value. There's your trade-off. Maclaurin series are simpler than Taylor series, but Maclaurin series are, by definition, centered at $x = 0$. If you're approximating a function value for an x far from 0, you'll have to use the slightly more complicated Taylor series.

Are you wondering where I am going with this metaphor? Wonder no longer. Maclaurin series are the fast-food approximators of the function world. (In my metaphor, regular functions equals home cooking and Maclaurin equals McDonald's.) We have already learned a method of approximating functions (remember linear approximation?) but Maclaurin series offer a much more accurate approximation to a function. All you have to do is learn one general formula, and suddenly you can approximate even complicated function values.

Mathematically, we say the *Maclaurin series* $\sum_{n=0}^{\infty} \dfrac{f^{(n)}(0)x^n}{n!}$ gives a very good approximation for the function values of $f(x)$, for x-values very close to 0. Believe it or not, you don't have to worry at all about the convergence of this series! In fact, when you work with Maclaurin series, you'll actually be generating *Maclaurin polynomials*, which are finite bits of the series.

Before we start generating these polynomials, let's make sure you understand the formula; it contains something weird—an exponent in parentheses. In case you don't know what $f^{(n)}(0)$ means, it is the nth derivative of $f(x)$ evaluated at 0. For example, $f^{(5)}(0)$ is the fifth derivative of $f(x)$ with 0 plugged in for x. This is handy notation. It's pretty obvious that writing the twelfth derivative of $f(x)$ as $f^{(12)}(x)$ is better than $f''''''''''''(x)$. (Although you could make a good argument that the latter notation is really in its "prime.")

Talk the Talk

The **Maclaurin series** $\sum_{n=0}^{\infty} \dfrac{f^{(n)}(0)x^n}{n!}$ gives a good approximation for values of the function $f(x)$, as long as you're approximating function values for x's close to $x = 0$. You won't use an infinite series to do the approximating, however; you'll use terms of the series up to a certain number n. The higher that n is, the closer your approximation will be to the actual function value.

Example 3: Use the fifth-degree Maclaurin polynomial for the function $f(x) = \sin x$ to approximate $\sin(0.1)$.

Solution: We need to write the terms of the Maclaurin series from $n = 0$ to $n = 5$:

$$\frac{f^{(0)}(0)x^0}{0!} + \frac{f'(0)x^1}{1!} + \frac{f''(0)x^2}{2!} + \cdots + \frac{f^{(5)}(0)x^5}{5!}$$

So, we'll need five derivatives of the function $f(x) = \sin x$ and we plug 0 into each one:

$$f(x) = \sin x \quad f(0) = \sin 0 = 0$$
$$f'(x) = \cos x \quad f'(0) = \cos 0 = 1$$
$$f''(x) = -\sin x \quad f''(0) = -\sin 0 = 0$$
$$f'''(x) = -\cos x \quad f'''(0) = -\cos 0 = -1$$
$$f^{(4)}(x) = \sin x \quad f^{(4)}(0) = \sin 0 = 0$$
$$f^{(5)}(x) = \cos x \quad f^{(5)}(0) = \cos 0 = 1$$

Kelley's Cautions

The first term in the series in Example 3 is bizarre: $\dfrac{f^{(0)}(0)x^0}{0!}$. There's no such thing as the "zeroth" derivative—that just means the original function (you derive it zero times). You may also wonder what the value of 0! is. Since the ordinary definition of factorial doesn't work with 0, we have to define its value separately: $0! = 1$.

Now, we need to plug these values into the series. In case you're confused about how that works, here's how you do it for the $n = 3$ term, $\dfrac{f'''(0)x^3}{3!}$. Since $f'''(0) = -1$ (according to our above work) and $3! = 3 \cdot 2 \cdot 1 = 6$, plug those values into the term to get $\dfrac{(-1)x^3}{6}$. Notice that the even-powered function derivatives will disappear as we plug into the function since you end up multiplying by 0 in each:

$$\frac{f^{(0)}(0)x^0}{0!} + \frac{f'(0)x^1}{1!} + \frac{f''(0)x^2}{2!} + \cdots + \frac{f^{(\ )}(0)x^5}{5!}$$

$$= 0 + \frac{1 \cdot x}{1} + 0 + \frac{(-1)x^3}{6} + 0 + \frac{1 \cdot x^5}{120}$$

$$= x - \frac{x^3}{6} + \frac{x^5}{120}$$

There you have it—the Maclaurin polynomial of degree 5 for $f(x) = \sin x$. If we plug $x = 0.1$ into the polynomial, we get our approximation:

$$\sin 0.1 \approx (0.1) - \frac{(0.1)^3}{6} + \frac{(0.1)^5}{120}$$

$$\approx .099833416667$$

The actual value of sin 0.1 is .099833416647. Wow! Talk about close!

Even for a hard-edged and jaded calculus expert like you, it's hard not to be surprised by just how accurate that approximation was. I can't approximate sin 0.1 to save my life. I barely even know how to start guessing at what its value would be. However, I got almost that exact value by plugging into a polynomial with *three terms*. If I had used $n = 7$, it would have been an even better approximation.

Take a look at how the graphs of the Maclaurin polynomials more closely resemble the graph of the sine function as the number of terms in the Maclaurin polynomial increases. The Maclaurin polynomial slowly shapes itself to exactly match the polynomial as you increase the number of polynomial terms. It reminds me of that movie *Single White Female*, where the crazy woman slowly adjusts every aspect of her life to match her roommate's with the eventual goal of killing her and taking over her life. I'm not saying that Maclaurin polynomials are crazy; just be careful ....

Figure 24.1 displays the graphs of the Maclaurin polynomials of degree 3, 5, and 7, respectively, compared to the graph of $f(x) = \sin x$. The higher the n, the better the polynomial approximates values of the sine function farther away from $x = 0$. If you could write the polynomial with the infinite number of terms suggested by $\sum\limits_{n=0}^{\infty} \dfrac{f^{(n)}(0)x}{n!}$, you'd get almost the exact graph of the sine function!

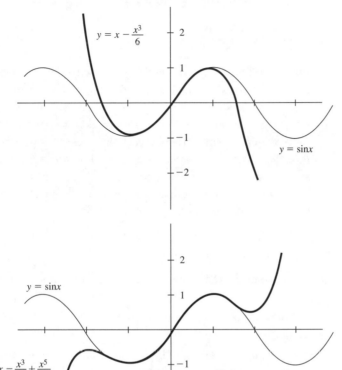

Figure 24.1

The more terms in the Maclaurin polynomial, the more closely its graph resembles the graph of sine.

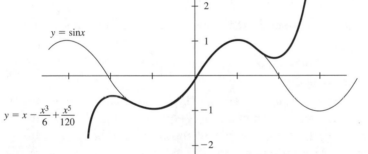

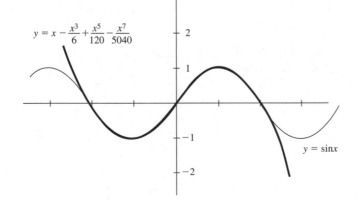

Taylor Series

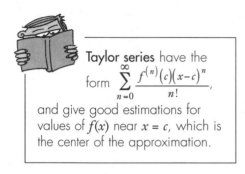

Taylor series have the form $\sum_{n=0}^{\infty} \frac{f^{(n)}(c)(x-c)^n}{n!}$, and give good estimations for values of $f(x)$ near $x = c$, which is the center of the approximation.

Once you know how Maclaurin series work, Taylor series are a piece of cake. Maclaurin series look so much like Taylor series because Maclaurin series actually are Taylor series centered at $x = 0$. Therefore, a Taylor series is just a more generic form of the Maclaurin series; it can be centered at any x-value, not just at x = 0. The goal of a Taylor series is the same as its predecessor—to approximate function values. Just like before, you're only guaranteed a good approximation if you stay close to the series' center.

Taylor series look almost identical to Maclaurin series; they have the form $\sum_{n=0}^{\infty} \frac{f^{(n)}(c)(x-c)^n}{n!}$, where c is the center of the series. Note the differences in the formulas:

- The derivatives in Taylor series are evaluated at $x = c$, the center of the approximation, not automatically at $x = 0$
- Rather than raising x to the n power, you raise the quantity $(x - c)$ to the n power

Other than that, the formulas are identical. Therefore, a Taylor polynomial is neither more nor less accurate than a Maclaurin polynomial—it is equally accurate but meant for a different purpose. If you are trying, for example, to approximate a function's value at $x = 10.1$, you'd need a lengthy Maclaurin polynomial, since 10.1 is so far away from Maclaurin's mandatory center of $x = 0$. However, a very small Taylor polynomial centered at $x = 10$ will give a great estimate.

Example 4: Approximate ln(2.1) using a third-degree Taylor polynomial for $f(x) = \ln x$ centered at $x = 2$.

Solution: In this case, $f(x) = \ln x$ and $c = 2$. We're going to need derivatives up to and including the third derivative, all evaluated at 2:

$$f(x) = \ln x \quad f(2) = \ln 2$$

$$f'(x) = \frac{1}{x} \quad f'(2) = \frac{1}{2}$$

$$f''(x) = -\frac{1}{x^2} \quad f''(2) = -\frac{1}{4}$$

$$f'''(x) = \frac{2}{x^3} \quad f'''(2) = \frac{1}{4}$$

Plug all of these puppies into the Taylor polynomial of degree 3:

$$f(2) + \frac{f'(2)(x-2)}{1!} + \frac{f''(2)(x-2)^2}{2!} + \frac{f'''(2)(x-2)^3}{3!}$$

$$= \ln 2 + \frac{1}{2} \cdot (x-2) + \left(-\frac{1}{4}\right)\frac{(x-2)^2}{2} + \frac{1}{4} \cdot \frac{(x-2)^3}{6}$$

$$= \ln 2 + \frac{x-2}{2} - \frac{(x-2)^2}{8} + \frac{(x-2)^3}{24}$$

To get your approximation of ln2.1, plug x = 2.1 into that polynomial:

$$\ln 2 + \frac{(2.1)-2}{2} - \frac{(2.1-2)^2}{8} + \frac{(2.1-2)^3}{24}$$

$$= \ln 2 + \frac{0.1}{2} - \frac{(0.1)^2}{8} + \frac{(0.1)^3}{24} \approx .74193885$$

The actual value of ln 2.1 is .74193734, so the approximation is incredibly accurate, considering how complex the function y = ln x is.

You've Got Problems

Problem 4: Approximate $\sqrt{4.2}$ using a second-degree Taylor polynomial for $f(x) = \sqrt{x}$ centered at x = 4.

The Least You Need to Know

◆ Power, Maclaurin, and Taylor series can be used to approximate function values close to their center.

◆ You find the radius of convergence for power series using the Ratio Test for absolute convergence.

◆ Even very compact Maclaurin polynomials can give terrific estimates of function values close to $x = 0$.

◆ Taylor series give estimates just as accurate as Maclaurin series, but you can center Taylor series at any x-value you choose.

Solutions to "You've Got Problems"

All of the answers to the problems that haunted you throughout the book are listed here, organized by chapter. All of the important steps are shown, unless the skill needed to complete a problem was already discussed in a previous chapter. For example, once you learn how to do u-substitution in Chapter 15. I no longer focus on its details if problems in subsequent chapters require u-substitution as a component of their answers. If I didn't do that, this appendix would be a book unto itself!

Chapter 2

1. $6x + 9y = 11$. Don't forget that $6x$ has to be positive to be in standard form. You may need to multiply everything by -1.

2. $2x - 3y = 6$. You can treat $(0,-2)$ as a point or use it as the y-intercept, so both forms work.

3. $\frac{3}{4}$. Remember that $\frac{-3}{-4} = \frac{3}{4}$.

4. $\frac{9y^4}{x^6}$. When you square everything, you get $9x^{-6}y^4$, and the negative exponent has to be moved.

5. $7xy(x - 3y^2)$. The greatest common factor is $7xy$, so divide it out of each term to get the factored form and write $7xy$ in front.

6. $(2x + 7)(4x^2 - 14x + 49)$. This is a sum of perfect cubes; $a = 2x$ and $b = 7$.

7. $x = 0,-4$. **Method one:** Factor out $3x$. **Method two:** First divide by 3 to get $x^2 + 4x = 0$. Half of 4 is 2, whose square, 4, should be added to both sides. **Method three:** $a = 3$, $b = 12$, and $c = 0$, since there is no constant term.

Chapter 3

1. 4. $f(43) = 7$; $g(7) = 64$; $h(64) = 4$.

2. Origin-symmetric. Plug in $-x$ for x and $-y$ for y to get $-y = \frac{-x^3}{|x|}$. Multiply both sides by -1, and you'll get the original function.

3. $\frac{1}{2}\left(\sqrt{2x+6}\right)^2 - 3 = \sqrt{2\left(\frac{1}{2}x^2 - 3\right) + 6} = x$ once simplifying is complete.

4. $h^{-1}(x) = \frac{3}{2}x - \frac{15}{2}$. After switching x and y, subtract 5 from both sides and then eliminate $\frac{2}{3}$ by multiplying each side of the equation by $\frac{3}{2}$.

5. $y = x^2 - 3x + 3$. Start by solving the x equation for t ($t = x - 1$), plug that into both t spots in the y equation, and simplify.

Chapter 4

1. 0. Simplify $\frac{14\pi}{4}$ to get $\frac{7\pi}{2}$. Subtract 2π $\left(\text{or } \frac{4\pi}{2}\right)$ to get $\frac{3\pi}{2}$; $\cos\frac{3\pi}{2} = 0$.

2. $\left(-\frac{1}{2}\right)^2 + \left(\frac{\sqrt{3}}{2}\right)^2 = \frac{1}{4} + \frac{3}{4} = 1$.

3. $\sin 2x \cos 2x$. Factor out the greatest common factor of $2\sin x \cos x$ to get $2\sin x \cos x(1 - 2\sin^2 x)$ and use double angle formulas to substitute in replacements for each factor.

4. $x = 0,\pi$. Substitute $2\sin x \cos x$ for $\sin 2x$ and factor to get $2\sin x(\cos x + 1) = 0$; solve each equation set equal to 0.

Chapter 5

1. $-\infty$. The graph decreases infinitely as you approach $x = -4$ from the left. You can also answer that no limit exists because the graph decreases infinitely—both methods of answering are equivalent.

2. Does not exist. The left-hand limit (–2) does not equal the right-hand limit (3), so no general limit exists.

3. 1. The left- and right-hand limits are both 1, so the general limit exists and is 1.

Chapter 6

1. (a) $-\frac{1}{\pi}$. Plug in π for each x to get $\frac{\cos \pi}{\pi}$; you know that $\cos\pi = -1$ from the unit circle.

   (b) $\frac{5}{3} \cdot \frac{(-2)^2 + 1}{(-2)^2 - 1} = \frac{4+1}{4-1}$.

2. (a) 13. Factor the numerator to get $(2x + 3)(x - 5)$; cancel the $(x - 5)$ terms and plug in $x = 5$ into $2x + 3$.

   (b) 3. The numerator is the difference of perfect cubes (remember the formula?), which factors to $(x - 1)(x^2 + x + 1)$; the $(x - 1)$ terms cancel, leaving only $x^2 + x + 1$; substitute $x = 1$ into that expression to get the answer.

3. (a) 4. Multiply numerator and denominator by $\sqrt{x + 6} + 2$ and cancel out resulting $(x + 2)$ terms to get $\lim_{x \to -2}\left(\sqrt{x+6} + 2\right)$; substitute in $x = -2$.

   (b) $\frac{\sqrt{5}-3}{2}$. Did I fool you? You don't use the conjugate method here, because substitution works; to get the answer, just plug in $x = 1$ for all x's (no simplifying can be done).

4. Factor to get $\frac{x(2x-1)(x-1)}{x(2x-1)(x+3)}$. The function is undefined at $x = 0$, $x = \frac{1}{2}$, and $x = -3$.

   Using the factoring method, $\lim_{x \to 0} g(x) = -\frac{1}{3}$ and $\lim_{x \to 1/2} g(x) = -\frac{1}{7}$, so holes exist on the graph for those values. However, no limit exists for $x = -3$, since substitution results in $-\frac{84}{0}$, indicating that $x = -3$ is a vertical asymptote.

5. (a) $\frac{2}{3}$. The degrees of the numerator and denominator are the same.

   (b) 0. The denominator has the higher degree; the fact that you're approaching $-\infty$ doesn't matter, since all rational functions possessing an infinite limit approach the same height as x approaches ∞ and $-\infty$.

6. e. Break into two limits to get $\lim_{x \to \infty} \frac{5}{x^3} + \lim_{x \to \infty}\left(1 + \frac{1}{x}\right)^x$; each of these is a separate special limit rule. The first limit is equal to 0 (by the third rule) and the other limit is equal to e (by the last rule), so the answer is $0 + e = e$.

Chapter 7

1. Discontinuous. From the function, it's clear that $g(1) = -2$, but using the factoring method, you get that $\lim\limits_{x \to 1} g(x) = 5$. Because these are unequal, g is discontinuous at $x = 1$.

2. $a = 12$. The function will approach a height of -6 when $x = -1$, because $2(-1)^2 + (-1)-7 = -6$. (Even though it technically doesn't reach that height, since the domain restriction is $x < -1$, -6 is still the left-hand limit as x approaches -1.) Therefore, $ax + 6 = -6$ when you plug in $x = -1$.

3. $x = 5$ (infinite discontinuity), $x = -5$ (point discontinuity). Factor to get $\dfrac{(x+5)(2x-5)}{(x+5)(x-5)}$; g increases infinitely at $x = 5$, whereas a limit exists for $x = -5$.

4. Since $g(1) = -2$ and $g(2) = 4$, we know that all values between -2 and 4 are outputs of g for some input between 1 and 2. Clearly, 0 is between -2 and 4, so the function has a height of 0 (causing an x-intercept) somewhere between $x = 1$ and $x = 2$.

Chapter 8

1. $g'(x) = 10x + 7$; $g'(-1) = -3$ First, calculate $g(x + \Delta x)$:

$$g(x + \Delta x) = 5(x + \Delta x)^2 + 7(x + \Delta x) - 6$$
$$= 5x^2 + 10x\Delta x + 5(\Delta x)^2 + 7x + 7\Delta x - 6$$

After plugging this into the difference quotient and simplifying, you get $\lim\limits_{\Delta x \to 0} \dfrac{10x\Delta x + 5(\Delta x)^2 + 7\Delta x}{\Delta x}$. Solve using the factoring method.

2. $\frac{1}{6}$. Since $h(8) = \sqrt{9} = 3$, the difference quotient is $\lim\limits_{x \to 8} \dfrac{\sqrt{x+1}-3}{x-8}$. Find this limit using the conjugate method.

Chapter 9

1. (a) $y' = 2x^2 + 6x - 6$. Here is the work behind the scenes:

$$y' = \left(\frac{2}{3} \cdot 3\right)x^{3-1} + (3 \cdot 2)x^{2-1} - (6 \cdot 1)x^{1-1} + 0$$

(b) $f'(x) = \dfrac{1}{3x^{2/3}} + \dfrac{2}{5x^{4/5}}$. Begin by writing the radical terms as fractional exponents and then apply the Power Rule:

$$f(x) = x^{1/3} + 2x^{1/5}$$

$$f'(x) = \left(1 \cdot \frac{1}{3}\right)x^{1/3-1} + \left(2 \cdot \frac{1}{5}\right)x^{1/5-1}$$

$$= \frac{1}{3}x^{-2/3} + \frac{2}{5}x^{-4/5}$$

2. To use the Power Rule, you must multiply to get $g(x) = 2x^2 + 7x - 4$ and derive that to get $g'(x) = 4x + 7$. Applying the Product Rule results in:

$$g'(x) = (2x - 1)(1) + (2)(x + 4)$$

$$= 2x - 1 + 2x + 8$$

$$= 4x + 7$$

3. Make sure to simplify carefully:

$$f'(x) = \frac{(x - 5)(12x^3 + 4x - 7) - (3x^4 + 2x^2 - 7x)(1)}{(x - 5)^2}$$

$$= \frac{(12x^4 - 60x^3 + 4x^2 - 27x + 35) - 3x^4 - 2x^2 + 7x}{x^2 - 10x + 25}$$

$$= \frac{9x^4 - 60x^3 + 2x^2 - 20x + 35}{x^2 - 10x + 25}$$

4. $10x(x^2 + 1)^4$. Here we have a function $(x^2 + 1)$ inside another function (x^5). In the Chain Rule formula, $f(x) = x^5$ and $g(x) = x^2 + 1$, since $f(g(x)) = (x^2 + 1)^5$. Therefore, you use the Power Rule to derive the outer function (while leaving $x^2 + 1$ alone) and then multiply by the derivative of $x^2 + 1$ to get $5(x^2 + 1)^4 \cdot (2x)$.

5. (a) 19. The instantaneous rate of change is synonymous with the derivative, so find $g'(4)$; since $g'(x) = 6x - 5$ because of the Power Rule, $g'(4) = 19$.

(b) 1. You'll need to find the slope of the secant line, so first get the points representing the x-values of -1 and 3 by plugging those x-values into the equation. Since $g(-1) = 14$ and $g(3) = 18$, the endpoints of the secant line are $(-1,14)$ and $(3,18)$. The secant slope will be $\frac{18-14}{3-(-1)} = \frac{4}{4} = 1$.

6. Begin by writing $\cot x$ as a quotient: $\cot x = \frac{\cos x}{\sin x}$.

$$\frac{d}{dx}(\cot x) = \frac{\sin x(-\sin x) - \cos x(\cos x)}{\sin^2 x}$$

$$= \frac{-\sin^2 x - \cos^2 x}{\sin^2 x}$$

Now, factor a –1 out of the numerator and use the Mama theorem to replace $\sin^2 x + \cos^2 x$ with 1:

$$\frac{d}{dx}(\cot x) = \frac{-\left(\sin^2 x + \cos^2 x\right)}{\sin^2 x}$$

$$= \frac{-1}{\sin^2 x}$$

$$= -\csc^2 x$$

Chapter 10

1. $y = 15x + 5$. The point of tangency is $(-1,-10)$ and since $g'(x) = 9x^2 - 2x + 4$, $g'(-1) = 15$. Point-slope form gives you $y - (-10) = 15(x - (-1))$, which you can put into slope-intercept form like I did, if you wish.

2. $\frac{2}{3}$. The derivative, with respect to x, is $4 + x\frac{dy}{dx} + y - 6y\frac{dy}{dx} = 0$. Solve this for $\frac{dy}{dx}$ to get $\frac{dy}{dx} = \frac{-4-y}{x-6y}$. To finish, plug in 3 for x and 2 for y and simplify.

3. 3. Evaluating $f^{-1}(6)$ is the same as solving $\sqrt{2x^3 - 18} = 6$. Square both sides to get $2x^3 - 18 = 36$, and solve for x by adding 18 to both sides, dividing both sides by 2, and then cube rooting both sides of the equation.

4. .0945. Remember that $\left(g^{-1}\right)'(-2) = \frac{1}{g'\left(g^{-1}(2)\right)}$ and $g^{-1}(2)$ is the solution to the equation $3x^5 + 4x^3 + 2x + 1 = -2$, which is $-.6749465398$. Therefore,

$$\left(g^{-1}\right)'(2) = \frac{1}{g'(-.6749465398)} = .0945 .$$

5. $\frac{dy}{dx} = \frac{\sec^2 t}{2} = \frac{1}{2}\sec^2 t$. This is the derivative of the y piece divided by the x piece's derivative. To get the second derivative, derive $\frac{dy}{dx}$ (with the Chain Rule) and divide by 2 (the original x equation derivative):

$$\frac{d^2 y}{dx^2} = \frac{\frac{1}{2} \cdot 2(\sec t) \cdot \sec t \tan t}{2} = \frac{1}{2}\sec^2 t \tan t$$

Chapter 11

1. First of all, $h'(x) = -2x + 6$. When you set that equal to 0 and solve, you get the critical number of $x = 3$. Choose numbers before and after 3 and plug them into the derivative—for example, $h'(2) = 2$ and $h'(4) = -2$. Since the derivative changes from positive to negative, the function changes from increasing to decreasing at $x = 3$, so the critical number represents a relative maximum.

2. Find the derivative: $g'(x) = 6x^2 - 7x - 3$; critical points are where this equals 0 (it is never undefined). So, factor to get $(3x + 1)(2x - 3)$, and critical numbers are $x = -\frac{1}{3}$ and $x = \frac{3}{2}$. Pick points and plug into the derivative and you get this wiggle graph:

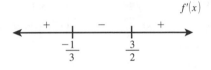

Because $f'(x)$ is positive on the intervals $\left(-\infty, -\frac{1}{3}\right)$ and $\left(\frac{3}{2}, \infty\right)$, f is increasing on those intervals as well.

3. Absolute max: 32; absolute min: –52. Because $g'(x) = 3x^2 + 8x + 5$, that factors to $(3x + 5)(x + 1)$, $x = -\frac{5}{3}$ and $x = -1$ are both critical numbers. A wiggle graph verifies that they are also relative extrema. So, test all four x-value candidates, including those and the endpoints: $f(-5) = -52$, $f\left(-\frac{5}{3}\right) \approx -3.852$, $f(-1) = -4$, and $f(2) = 32$.

4. $\left(0, \frac{\pi}{2}\right)$ and $\left(\frac{3\pi}{2}, 2\pi\right)$. If $f(x) = \cos x$, then $f'(x) = -\sin x$ and $f''(x) = -\cos x$. The second derivative wiggle graph for $[0, 2\pi]$ looks like this:

... and f will be concave down wherever this graph is negative.

Chapter 12

1. $t = 4$ and $t = 8.196$ seconds. This question is asking, "When is the position equal to –30?" To answer it, use some form of technology to solve the equation

$\frac{1}{2}t^3 - 5t^2 + 3t + 6 = -30$. Again, I usually set it equal to 0 and find the x-intercepts (i.e., solve the equation $\frac{1}{2}t^3 - 5t^2 + 3t + 36 = 0$). Negative answers make no sense and should be discarded. (Negative time is nonsensical.)

2. The correct order is: the average velocity, the velocity at $t = 7$, and lastly the speed at $t = 3$. The average velocity is the slope connecting the points $(2,-4)$ and $(6,-48)$: $\frac{-48-(-4)}{6-2} = \frac{-44}{4} = -11$ in/sec. The velocity at $t = 7$ is $v(7) = s'(7) = 6.5$ in/sec. The speed at $t = 3$ is the absolute value of the velocity there: $|s'(3)| = |-13.5| = 13.5$ in/sec.

3. $t = 3$ seconds. Since $s''(t) = 3t - 10$, the answer is the solution to the equation $3t - 10 = -1$.

4. 585.204 meters. The position equation will be $s(t) = -4.9t^2 + 100t + 75$. The highest point reached by the cannonball is the relative maximum of the position equation. Since $s'(t) = -9.8t + 100$, $t = \frac{-100}{-9.8} = 10.204$ is the time the ball reaches this height (verify it's a max using the Second Derivative Test if you like—$s''(t)$ is always negative). Thus, the maximum height of the cannonball is $s(10.204)$, which is approximately 585.204 meters.

Chapter 13

1. 0. Since x^{-2} has a negative power, move it to the denominator: $\lim_{x \to \infty} \frac{\ln x}{x^2}$. Substitution results in $\frac{\infty}{\infty}$, so apply L'Hôpital's Rule (and remember that the derivative of $\ln x$ is $\frac{1}{x}$: $\lim_{x \to \infty} \frac{\frac{1}{x}}{2x}$. This can be rewritten as $\lim_{x \to \infty} \frac{1}{2x^2}$. Substitution now results in 1 divided by a giant number, which is basically 0, according to the third of our special limit theorems from Chapter 6.

2. $x = \frac{1}{2}$. Since $g\left(\frac{1}{4}\right) = 4$ and $g(1) = 1$, the secant slope is $\frac{4-1}{\frac{1}{4}-1} = \frac{3}{-\frac{3}{4}} = -4$. The Power Rule tells you that $g'(x) = -\frac{1}{x^2}$. Thus, a solution to $-\frac{1}{x^2} = -4$ is the value for c guaranteed by the Mean Value Theorem:

$$x^2 = \frac{1}{4}$$

$$x = \pm\frac{1}{2}$$

Only $x = \frac{1}{2}$ falls in the interval $\left[\frac{1}{4},1\right]$, so discard the other answer.

3. 2.857 in²/week. We know, from the problem, that $\frac{dV}{dt} = 5$ if V represents volume. Let's call S surface area; we want to find $\frac{dS}{dt}$. A good equation involving it is $S = 6l^2$, where l is the length of a side. Think about it—the surface area of a cube is six squares, each having area l^2. So, derive that formula to get $\frac{dS}{dt} = 12l \cdot \frac{dl}{dt}$. We know that $l = 7$, but what is $\frac{dl}{dt}$? To find it, we have to use the given knowledge about $\frac{dV}{dt}$, so we need an equation containing V. The volume of a cube of side l is $V = l^3$, so let's derive that baby to get $\frac{dV}{dt} : \frac{dV}{dt} = 3l^2 \cdot \frac{dl}{dt}$. We know that $\frac{dV}{dt} = 5$ and $l = 7$, so plug it into this new equation to get $5 = 3 \cdot 7^2 \cdot \frac{dl}{dt}$, so $\frac{dl}{dt} = \frac{5}{147}$. Now that we finally know what $\frac{dl}{dt}$ is, plug it back into our $\frac{dS}{dt}$ equation to solve for $\frac{dS}{dt} : \frac{dS}{dt} = 12(7) \cdot \frac{5}{147} \approx 2.857$ in²/week.

4. $-\frac{9}{8}$. You want to optimize the product, whose equation is $P = xy$, where x and y are the numbers in question. You know that $y = 2x - 3$, so $P = x(2x - 3) = 2x^2 - 3x$. So, $P' = 4x - 3$, and the wiggle graph of P' is …

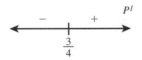

So, one of the numbers is $\frac{3}{4}$, and the other is $y = 2 \cdot \frac{3}{4} - 3 = -\frac{3}{2}$. Remember, you're asked for the optimal product, so the answer is $-\frac{9}{8}$.

Chapter 14

1. The width of all the rectangles will be $\Delta x = \frac{\frac{3\pi}{2} - \frac{\pi}{2}}{4} = \frac{\pi}{4}$. The left-hand sum will be (I factored out the $\frac{\pi}{4}$ width from each term to make the answers easier to read):

$$\frac{\pi}{4}\left(-\cos\frac{\pi}{2} - \cos\frac{3\pi}{4} - \cos\pi - \cos\frac{5\pi}{4}\right) \approx 1.89612$$

The right-hand sum will be ...

$$\frac{\pi}{4}\left(-\cos\frac{3\pi}{4} - \cos\pi - \cos\frac{5\pi}{4} - \cos\frac{3\pi}{2}\right) \approx 1.89612$$

The midpoint sum is ...

$$\frac{\pi}{4}\left(-\cos\frac{5\pi}{8} - \cos\frac{7\pi}{8} - \cos\frac{9\pi}{8} - \cos\frac{11\pi}{8}\right) \approx 2.05234$$

2. 1.89612. Each trapezoid has width $\Delta x = \frac{\pi-0}{4} = \frac{\pi}{4}$, so according to the Trapezoidal Rule:

$$\frac{\pi-0}{2(4)}\left(\sin 0 + 2\sin\frac{\pi}{4} + 2\sin\frac{\pi}{2} + 2\sin\frac{3\pi}{4} + \sin\pi\right)$$

$$= \frac{\pi}{8}\left(0 + 2\cdot\frac{\sqrt{2}}{2} + 2\cdot 1 + 2\cdot\frac{\sqrt{2}}{2} + 0\right)$$

$$= \frac{\pi}{8}\left(2\sqrt{2} + 2\right) \approx 1.89612$$

3. 1.622. Each subinterval has the width of $\Delta x = \frac{5-1}{4} = 1$; apply the Simpson's Rule formula:

$$\frac{5-1}{3\cdot 4}\left(f(1) + 4f(2) + 2f(3) + 4f(4) + f(5)\right)$$

$$= \frac{4}{12}\left(1 + 4\cdot\frac{1}{2} + 2\cdot\frac{1}{3} + 4\cdot\frac{1}{4} + \frac{1}{5}\right)$$

$$= \frac{1}{3}\left(4 + \frac{13}{15}\right) = \frac{73}{45} \approx 1.622$$

Chapter 15

1. $\frac{2}{5}x^5 + \frac{1}{12}x^4 + \frac{2}{3}x^{3/2} + C$. Start by writing each term as a separate integral with a coefficient and power: $\int 2x^4 dx + \int\frac{1}{3}x^3 dx + \int x^{1/2} dx$. Now, factor out the coefficients to get $2\int x^4 dx + \frac{1}{3}\int x^3 dx + \int x^{1/2} dx$. Finally, apply the Power Rule for integrals and simplify:

$$2 \cdot \frac{x^5}{5} + \frac{1}{3} \cdot \frac{x^4}{4} + \frac{x^{3/2}}{\dfrac{3}{2}} + C$$

$$\frac{2}{5}x^5 + \frac{1}{12}x^4 + \frac{2}{3}x^{3/2} + C$$

Don't get confused when adding 1 to the fractional power: $1 + \frac{1}{2} = \frac{3}{2}$.

2. –2. The integral of $\cos x$ is $\sin x$ (not $-\sin x$, which is the derivative of $\cos x$). So, plug the limits of integration into the integral in the correct order:

   $\sin \frac{3\pi}{2} - \sin \frac{\pi}{2} = -1 - 1 = -2$. This is the area between the graph of $y = \cos x$ and the x-axis. As you can see by the graph of cosine …

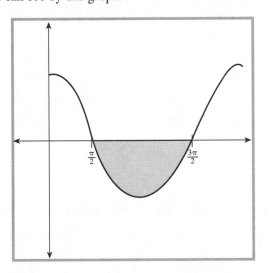

   … that area is below the x-axis, so the definite integral will be negative.

3. **Part one:** Start by evaluating the definite integral (remember the integral of e^t is e^t: $\frac{d}{dx}\left(e^t \Big|_1^{\tan x}\right) = \frac{d}{dx}\left(e^{\tan x} - e\right)$. Now derive; since e is a constant (there is no x exponent) its derivative is 0: $e^{\tan x} \cdot \sec^2 x$. (You use the Chain Rule, at first leaving the exponent alone and then multiplying by its derivative.)

   Part two: Because you are deriving with respect to the variable in the upper bound (and the lower bound is a constant), plug the upper bound into the function and multiply by the upper bound's derivative: $e^{\tan x} \cdot \sec^2 x$.

4. $\frac{1}{2}$. If you rewrite the integral as $\int_0^{\pi/4} \sec x \cdot \sec x \tan x \, dx$, the u is easier to find.

   Because the derivative of $\sec x$ is found in the problem, you set $u = \sec x$ and derive to get $du = \sec x \tan x \, dx$. Now, replace $\sec x$ with u, $\sec x \tan x \, dx$ with du, and also replace the boundaries. Plug them each in for x in the equation $u = \sec x$ to get their

u equivalents. Because $\sec \frac{\pi}{4} = \frac{2}{\sqrt{2}}$ and $\sec 0 = 1$, these are the new boundaries. With all these changes you get $\int_{1}^{2/\sqrt{2}} u \, du$, which is quite simple to solve:

$$\left. \left(\frac{u^2}{2} \right) \right|_{1}^{2/\sqrt{2}} = \frac{\frac{4}{2}}{2} - \frac{1}{2} = 1 - \frac{1}{2} = \frac{1}{2}$$

By the way, you could have also set $u = \tan x$ to solve the problem (resulting in $du = \sec^2 x \, dx$)—you'll get the same answer either way. Some would argue this substitution is even easier!

Chapter 16

1. $\frac{1}{12}$. These curves intersect at $x = 0$ and $x = 1$ (which you deduce by setting $x^2 = x^3$ and solving for x so those x-values bound the area the functions enclose. The graph of x^2 is above x^3 on that interval, so the area will be $\int_{0}^{1} \left(x^2 - x^3 \right) dx$, which equals:

$$\left. \left(\frac{x^3}{3} - \frac{x^4}{4} \right) \right|_{0}^{1} = \left(\frac{1^3}{3} - \frac{1^4}{4} \right) - 0 = \frac{1}{12}$$

2. .107. According to the Mean Value Theorem for Integration, we know that $\left(100 - 1 \right) \cdot f\left(c \right) = \int_{1}^{100} \frac{\ln x}{x} \, dx$. To integrate, you have to use a u-substitution with $u = \ln x$ and $du = \frac{1}{x} dx$:

$$99 f\left(c \right) = \int_{\ln 1}^{\ln 100} u \, du$$
$$99 f\left(c \right) = \left. \frac{u^2}{2} \right|_{\ln 1}^{\ln 100}$$
$$99 f\left(c \right) = \frac{\left(\ln 100 \right)^2}{2} - 0$$
$$f\left(c \right) \approx .107$$

3. 71,000 miles. Distance-traveled problems require us to know the *velocity* equation, so differentiate the position equation to get $v(t) = 3t^2 - 4t - 4$. Now create a wiggle graph of $v(t)$:

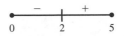

The ship changes direction (i.e., starts heading away from earth) at $t = 2$, so you have to use to integrals for velocity—one for **[0,2]** and one for **[2,5]**. Since the integral on **[0,2]** will be negative, you need to multiply it by -1. Total distance will be …

$$-\int_0^2 \left(3t^2 - 4t - 4\right)dt + \int_2^5 \left(3t^2 - 4t - 4\right)dt$$
$$= -\left(-8\right) + 63 = 71$$

4. $g(4\pi) = 0; g'(4\pi) = \tfrac{1}{2}.$

(a) To calculate $g(4\pi)$, plug it into the integral and evaluate it. You'll have to use *u*-substitution to integrate $\cos 2t$:

$$g(4\pi) = \int_{-\pi}^{(4\pi)/2} \cos 2t \; dt$$
$$= \frac{1}{2}\left(\sin u\right)\Big|_{-2\pi}^{4\pi} = 0$$

(b) Begin by finding $g'(x)$ using the Fundamental Theorem part two (plug $\tfrac{x}{2}$ into t and multiply by $\tfrac{1}{2}$). Then evaluate the derivative normally:

$$g'(x) = \cos\left(2 \cdot \frac{x}{2}\right) \cdot \frac{1}{2} = \frac{1}{2}\cos x$$
$$g'(4\pi) = \frac{1}{2}\cos(4\pi) = \frac{1}{2} \cdot 1 = \frac{1}{2}$$

Chapter 17

1. $-\ln\!\left|\cos x\right| + x + C$. If you write each fraction separately with $\cos x$ in the denominator of each, you get:

$$\int \frac{\sin x}{\cos x}\,dx + \int \frac{\cos x}{\cos x}\,dx$$
$$= \int \tan x \; dx + \int 1 \; dx$$

You memorized the integral of tangent, and the integral of 1 couldn't be easier.

2. $x + 2\ln\!\left|2x - 3\right| + C$. (a) Set $u = 2x - 3$; it gives you $\frac{du}{2} = dx$. In addition, solve the u equation for x to get $x = \frac{u+3}{2}$. Substitute all of these into the original integral and solve:

$$\frac{1}{2}\int \frac{2\left(\dfrac{u+3}{2}\right)+1}{u}\, du$$

$$= \frac{1}{2}\int \frac{u+4}{u}\, du$$

$$= \frac{1}{2}\left(\int du + 4\int \frac{1}{u}\, du\right)$$

$$= \frac{1}{2}\left(u + 4\ln|u| + C\right)$$

$$= \frac{1}{2}\left(2x-3\right) + 2\ln|2x-3| + C$$

$$= x + 2\ln|2x-3| + C$$

(b) Using long division, rewrite $\frac{2x+1}{2x-3}$ as $1 + \frac{4}{2x-3}$; you get the same answer as in part (a).

3. $\frac{3}{2\sqrt{7}}\arctan\left(\frac{x^2}{\sqrt{7}}\right) + C$. The denominator is a number plus a function squared, i.e., $(x^2)^2$. Pull the 3 out of the integral, set $a = \sqrt{7}$ (since $a^2 = \left(\sqrt{7}\right)^2 = 7$), and set $u = x^2$. Using u-substitution, $du = 2x\, dx$, so $\frac{du}{2} = x\, dx$. Using all these pieces, rewrite the integral and solve:

$$\frac{3}{2}\int \frac{du}{a^2 + u^2} = \frac{3}{2}\left(\frac{1}{\sqrt{7}}\arctan\left(\frac{x^2}{\sqrt{7}}\right)\right) + C$$

If radicals in the denominator bother you, feel free to rationalize the denominators, but the answer is acceptable as it is.

4. $\operatorname{arcsec}\left(\frac{x-3}{2}\right) + C$. Pull out the coefficient of 2 and complete the square inside the radical. The $(x-3)$ outside the radical and the order of subtraction in the radical suggests the arcsecant formula (arcsine usually has nothing outside the radical sign in the denominator):

$$2\cdot\int \frac{dx}{(x-3)\sqrt{x^2-6x+9-9+5}} = 2\cdot\int \frac{dx}{(x-3)\sqrt{(x-3)^2-4}}$$

Set $u = x - 3$ (so $du = dx$) and $a = 2$. When you do, you get the arcsecant formula exactly:

$$2\cdot\int \frac{du}{u\sqrt{u^2-a^2}}\, dx = 2\cdot\frac{1}{2}\operatorname{arcsec}\left(\frac{x-3}{2}\right) + C$$

Chapter 18

1. $x^2 e^x - 2xe^x + 2e^x + C$. To begin, set $u = x^2$ and $dv = e^x\, dx$, so $du = 2xdx$ and $v = e^x$. When you plug into the parts formula, you get $x^2 e^x - 2\int xe^x dx$. Use parts to integrate again, this time with $u = x$ and $dv = e^x\, dx$, to get $\int xe^x dx = xe^x - \int e^x dx$, which equals $xe^x - e^x$. Now that you know what $\int xe^x dx$ equals, plug it into our original parts formula: $x^2 e^x - 2(xe^x - e^x) + C$.

2. Using the same u and dv terms, you should get this table:

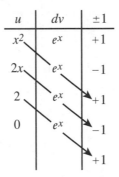

u	dv	± 1
x^2	e^x	$+1$
$2x$	e^x	-1
2	e^x	$+1$
0	e^x	-1
		$+1$

Therefore, your answer is $x^2 e^x - 2xe^x - 2e^x + C$.

3. $-\ln|x + 3| + \ln|x + 1| + C$. We can apply partial fractions since the denominator is factorable:

$$\frac{2}{x^2 + 4x + 3} = \frac{A}{x + 3} + \frac{B}{x + 1}$$
$$2 = A(x + 1) + B(x + 3)$$
$$2 = Ax + A + Bx + 3B$$
$$2 = x(A + B) + (A + 3B)$$

Because there is no x term on the left, $A + B = 0$. Because the constant on the left is 2, $A + 3B = 2$. This gives you two equations with two unknown variables—solve this system of equations (just as you did in elementary algebra) to get $A = -1$ and $B = 1$. Finish by substituting in the values:

$$\int \frac{2}{x^2 + 4x + 3}\, dx = \int \frac{-1}{x + 3}\, dx + \int \frac{1}{x + 1}\, dx$$
$$= -\ln|x + 3| + \ln|x + 1| + C$$

4. 2. The integral is improper because of its infinite upper limit of integration. Replace the troublesome limit with a and let a approach ∞. To integrate, rewrite the fraction as $x^{-3/2}$ and apply the Power Rule for Integration:

$$\lim_{a \to \infty} \left(\int_1^a x^{-3/2} \, dx \right)$$

$$= \lim_{a \to \infty} \left(-2x^{-1/2} \Big|_1^a \right)$$

$$= \lim_{a \to \infty} \left(\frac{-2}{\sqrt{x}} \Big|_1^a \right)$$

$$= \lim_{a \to \infty} \left(\frac{-2}{\sqrt{a}} - \left(\frac{-2}{\sqrt{1}} \right) \right) = 2$$

As a gets infinitely large, the denominator of $\frac{-2}{\sqrt{a}}$ will get huge, making the fraction basically equal to 0 (remember the special limit rules at the end of Chapter 6?).

Chapter 19

1. $\frac{81\pi}{2}$. Begin by drawing a picture of the situation:

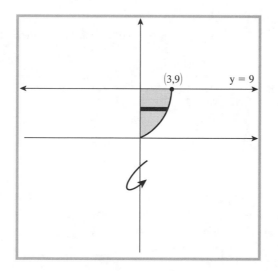

In this problem, the radius of rotation is horizontal—remember it must be perpendicular to the rotational axis, which is the y-axis here. The presence of a horizontal radius of rotation indicates that the function must contain y's, not x's, so solve the equation for x to accomplish this: $x = \pm\sqrt{y}$; we can ignore the case of $-\sqrt{y}$ since we're only in the first quadrant. The radius of rotation's length will be the right boundary minus the left boundary, so $r(y) = \sqrt{y} - 0$. Plug this into the formula for the disk method (using y-boundaries, since everything must be in terms of y):

$$V = \pi\int_0^9\left(\sqrt{y}\right)^2 dy$$
$$= \pi\int_0^9 y\,dy$$
$$= \frac{81\pi}{2}$$

2. $\frac{25\pi}{21}$. Remember that the rotational radii must extend from the axis of rotation (which is $y = -1$ in this problem), *not* always from the x-axis:

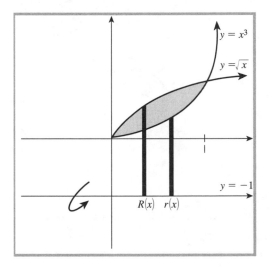

Therefore, $R(x) = \sqrt{x} - (-1) = \sqrt{x} + 1$ and $r(x) = x^3 - (-1) = x^3 + 1$ (subtract the bottom boundary (–1) from the top boundary for each). To get the correct volume, plug into the washer method. Use boundaries of 0 and 1 since they mark the x-values of the graphs' intersection points:

$$\pi\int_0^1\left(\left(\sqrt{x}+1\right)^2-\left(x^3+1\right)^2\right)dx$$
$$=\pi\int_0^1\left(x+2\sqrt{x}+1-x^6-2x^3-1\right)dx$$
$$=\pi\int_0^1\left(-x^6-2x^3+x+2x^{1/2}\right)dx$$
$$=\frac{25\pi}{21}$$

3. $\frac{9\pi}{35}$. If we're going to use the shell method, the radius involved must be parallel to the x-axis, so it must be horizontal; this means everything must be in terms of y. Solve both of your equations for x to get $x=y^2$ and $x=\sqrt[3]{y}$:

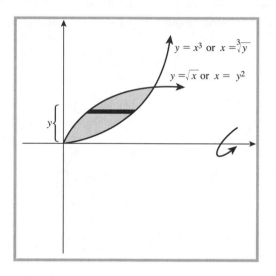

The radius is y units above the origin, so $d(y)=y$, while the length of the radius is the right equation minus the left equation: $h(y)=\sqrt[3]{y}-y^2$. The equations intersect at y values of 0 and 1. Plug everything into the shell method:

$$2\pi\int_0^1 y\left(\sqrt[3]{y}-y^2\right)dy$$
$$=2\pi\int_0^1\left(y^{4/3}-y^3\right)dy$$
$$=\frac{5\pi}{14}$$

4. $g(x) = x^3$. Use the arc length formula to find the length of each separately:

$$\int_0^2 \sqrt{1 + \left(f'(x)\right)^2}\, dx \qquad \int_0^2 \sqrt{1 + \left(g'(x)\right)^2}\, dx$$
$$= \int_0^2 \sqrt{1 + \left(2x\right)^2}\, dx \qquad = \int_0^2 \sqrt{1 + \left(3x^2\right)^2}\, dx$$
$$= \int_0^2 \sqrt{1 + 4x^2}\, dx \qquad = \int_0^2 \sqrt{1 + 9x^4}\, dx$$
$$\approx 4.6468 \qquad\qquad \approx 8.6303$$

The cubic graph is steeper, so it covers more ground during the same x-interval.

5. 8.2682. Since $\frac{dx}{dt} = 1$ and $\frac{dy}{dt} = 2t$, the arc length will be ...

$$\int_1^3 \sqrt{1^2 + \left(2t\right)^2}\, dt$$
$$= \int_1^3 \sqrt{1 + 4t^2}\, dt$$
$$\approx 8.268$$

Chapter 20

1. $y = \arcsin\left(\frac{1}{2}\ln\left|x^2 - 1\right| + C\right)$. Divide both sides by $(x^2 - 1)$ and multiply both sides by $\cos y$ to get:

$$\cos y\, dy = \frac{x\, dx}{x^2 - 1}$$

Integrate both sides (use u-substitution for the right side):

$$\sin y = \frac{1}{2}\ln\left|x^2 - 1\right| + C$$

Finally, solve for y by taking the arcsine of both sides (i.e., cancel out sine with its inverse function).

2. Integrate the acceleration function to find velocity. Since you know that $v(0) = -2$, substitute these values once you've integrated:

$$v(t) = \int a(t)dt = t^2 + 5t + \cos t + C$$
$$v(0) = 0^2 + 5 \cdot 0 + \cos 0 + C = -2$$
$$1 + C = -2$$
$$C = -3$$

Therefore, $v(t) = t^2 + 5t + \cos t - 3$. Integrate this to get the position function, this time using the fact that $s(0) = 5$ to find the C that results:

$$s(t) = \int \left(t^2 + 5t + \cos t - 3 \right) dt$$

$$= \frac{t^3}{3} + \frac{5t^2}{2} + \sin t - 3t + C$$

$$s(0) = \frac{0^3}{3} + \frac{5 \cdot 0^2}{2} + \sin 0 - 3 \cdot 0 + C = 5$$

$$C = 5$$

The final position equation is $s(t) = \dfrac{t^3}{3} + \dfrac{5t^2}{2} + \sin t - 3t + 5$.

3. 31.4 days. First things first; we need to calculate k. Our initial amount is 15,000, so that will equal N. After $t = 3.82$ days, 7,500 grams remain, so we can plug into our exponential decay equation:

$$y = Ne^{kt}$$

$$7,500 = 15,000e^{3.82k}$$

$$\frac{1}{2} = e^{3.82k}$$

$$\frac{\ln\left(\frac{1}{2}\right)}{3.82} = k$$

$$k \approx -.181452$$

Thus, our model for exponential decay is $y = 15,000e^{-.181452t}$. Set it equal to 50 and solve for t to resolve our dilemma:

$$y = 15,000e^{-.181452t}$$

$$50 = 15,000e^{-.181452t}$$

$$\frac{1}{300} = e^{-.181452t}$$

$$t = \frac{\ln\left(\frac{1}{300}\right)}{-.181452} \approx 31.4341 \text{ days}$$

Chapter 21

1. 1.08715. The slope of the tangent line to $f(x) = \arctan x$ is $f'(x) = \frac{1}{1+x^2}$; therefore, the slope of our linear approximation will be $f'(2) = \frac{1}{1+2^2} = \frac{1}{5}$. The point of tangency is $(2, \arctan 2)$. That gives us a linear approximation of …

$$y - \arctan 2 = \frac{1}{5}(x - 2)$$

$$y = \frac{1}{5}x - \frac{2}{5} + \arctan 2$$

Therefore, arctan1.9 is approximately …

$$\frac{1}{5}(1.9) - \frac{2}{5} + \arctan 2 \approx 1.08715$$

This is pretty close to the actual value; arctan(1.9) = 1.08632.

2. The slope field spirals counterclockwise; the specific solution to the differential equation passing through (0,1) should look like the darkened graph:

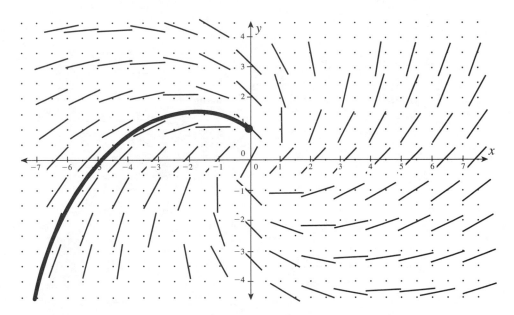

Determine the value of the slopes by plugging into the differential equation. For example, the slope of the segment at point (2,–1) will be …

$$\frac{dy}{dx} = \frac{x + y}{x - y} = \frac{2 - 1}{2 - (-1)} = \frac{1}{3}$$

3. $\left(-\frac{1}{2}, \frac{11}{3}\right)$. The slope is negative, so we'll be going down and to the right from our original point. We're traveling a distance of $\Delta x = \frac{1}{2}$ from our original point, so use that and the given slope to calculate Δy:

$$m = \frac{\Delta y}{\Delta x}$$

$$-\frac{2}{3} = \frac{\Delta y}{\frac{1}{2}}$$

$$\Delta y = -\frac{2}{3} \cdot \frac{1}{2} = -\frac{1}{3}$$

So, we should go right $\frac{1}{2}$ and down $-\frac{1}{3}$ from our original point of $(-1, 4)$ to stay on the line. Make those adjustments to the coordinate to get the answer:

$$\left(-1 + \frac{1}{2}, 4 - \frac{1}{3}\right)$$

$$= \left(-\frac{1}{2}, \frac{11}{3}\right)$$

4. $y(1) \approx \frac{16}{27}$. Here are all three steps:

Step one: $\frac{dy}{dx} = 2x - y = 2(0) - 0 = 0$; knowing $x = \frac{1}{3}$, find Δy:

$$0 = \frac{\Delta y}{\frac{1}{3}}$$

$$\Delta y = 0$$

This gives you a new point of $\left(0 + \frac{1}{3}, 0 + 0\right) = \left(\frac{1}{3}, 0\right)$.

Step two: $\frac{dy}{dx} = 2x - y = 2\left(\frac{1}{3}\right) - 0 = \frac{2}{3}$; knowing $x = \frac{1}{3}$, find Δy:

$$\frac{2}{3} = \frac{\Delta y}{\frac{1}{3}}$$

$$\Delta y = \frac{2}{3} \cdot \frac{1}{3} = \frac{2}{9}$$

This gives you a new point of $\left(\frac{1}{3}+\frac{1}{3}, 0+\frac{2}{9}\right) = \left(\frac{2}{3}, \frac{2}{9}\right)$.

Step three: $\frac{dy}{dx} = 2x - y = 2\left(\frac{2}{3}\right) - \frac{2}{9} = \frac{10}{9}$; knowing that $x = \frac{1}{3}$, find Δy:

$$\frac{10}{9} = \frac{\Delta y}{\frac{1}{3}}$$

$$\Delta y = \frac{10}{9} \cdot \frac{1}{3} = \frac{10}{27}$$

This gives you a new point of $\left(\frac{2}{3}+\frac{1}{3}, \frac{2}{9}+\frac{10}{27}\right) = \left(1, \frac{16}{27}\right)$.

Chapter 22

1. Divergent. The sequence converges if $\lim\limits_{n \to \infty} \frac{5n^3}{\ln n^2}$ exists. To evaluate the limit, you should use L'Hôpital's Rule:

$$\lim_{n \to \infty} \frac{15n^2}{\frac{1}{n^2} \cdot 2n} = \lim_{n \to \infty} \frac{15n^2}{\frac{2}{n}} = \lim_{n \to \infty} \frac{15}{2} n^3 = \infty$$

Because $\frac{15}{2}n^3$ increases without bound as n approaches infinity, the limit does not exist, making the sequence $\left\{\frac{5n^3}{\ln n^2}\right\}$ divergent.

2. Divergent. According to the nth term divergence test, $\sum\limits_{n=0}^{\infty} \sin n$ must diverge since $\lim\limits_{n \to \infty} \sin n \neq 0$. The sine function does not approach any fixed value as n increases—it oscillates infinitely between the values of 1 and –1.

3. Converges to a sum of 12. This is a geometric series with $a = 4$ and $r = \frac{2}{3}$. Because $0 < \left|\frac{2}{3}\right| < 1$, the series converges to sum:

$$\frac{a}{1-r} = \frac{4}{1-\frac{2}{3}} = \frac{4}{\frac{1}{3}} = 12$$

4. Convergent. You can pull out the constant and rewrite the negative exponent to get the p-series $\frac{6}{5} \cdot \sum\limits_{n=1}^{\infty} \frac{1}{n^3}$. Since $p = 3$ and $3 > 1$, the series converges.

5. 1. If you write out the expansion, you can easily tell that this is a telescoping series:

$$\left(1-\frac{1}{2}\right)+\left(\frac{1}{2}-\frac{1}{3}\right)+\left(\frac{1}{3}-\frac{1}{4}\right)+\left(\frac{1}{4}-\frac{1}{5}\right)+\cdots$$

In fact, all the terms of the series cancel out except the 1.

Chapter 23

1. Divergent. If the integral $\int_1^\infty \frac{\ln n}{n}dn$ (which can be solved via u-substitution with $u = \ln n$) diverges, so will the original series:

$$\lim_{a\to\infty}\left(\int_1^a \frac{\ln n}{n}dn\right)$$

$$=\lim_{a\to\infty}\left(\frac{u^2}{2}\right)\Big|_0^{\ln a}$$

$$=\lim_{a\to\infty}\left(\frac{(\ln a)^2}{2}\right)=\infty$$

2. Convergent. If it weren't for the 3 in the denominator, you'd have a p-series, so let's compare the given series to the convergent p-series $\sum_{n=1}^{\infty}\frac{1}{n^4}$. Notice that $\frac{1}{n^4+3}\leq\frac{1}{n^4}$ for all n, since the $+3$ in the denominator will cause the fraction to have a lesser value. (Adding in the numerator makes a fraction greater, and adding in the denominator does the opposite.) Therefore, $\sum_{n=1}^{\infty}\frac{1}{n^4+3}$ must be convergent according to the Comparison Test, because it is smaller than a convergent p-series.

3. Convergent. Since both parts of the fraction contain n raised to a power, the Limit Comparison Test is a great idea. A good comparison series will be $\sum_{n=1}^{\infty}\frac{n^{1/3}}{n^2}$, which can be simplified as $\sum_{n=1}^{\infty}\frac{1}{n^{5/3}}$. Now, calculate the limit of the quotient of the two series:

$$\lim_{n\to\infty}\frac{n^{1/3}}{n^2+1}\cdot\frac{n^{5/3}}{1}$$

$$\lim_{n\to\infty}\frac{n^2}{n^2+1}=1$$

Because the limit is positive and finite, and $\displaystyle\sum_{n=1}^{\infty}\frac{1}{n^{5/3}}$ is a convergent p-series, then $\displaystyle\sum_{n=1}^{\infty}\frac{\sqrt[3]{n}}{n^2+1}$ also converges.

4. Convergent. Applying the Ratio Test, you get:

$$\lim_{n\to\infty}\frac{(n+1)3^{n+1}}{(n+1)!}\cdot\frac{n!}{n\cdot3^n}$$

$$=\lim_{n\to\infty}\frac{(n+1)\cdot n!\cdot3^n\cdot3}{(n+1)\cdot n!\cdot n\cdot3^n}$$

$$=\lim_{n\to\infty}\frac{3}{n}=0$$

Because $0<1$, the series converges.

5. Convergent. You can rewrite this series to get $\displaystyle\sum_{n=1}^{\infty}\left(\frac{3^2}{n}\right)^n$ or $\displaystyle\sum_{n=1}^{\infty}\left(\frac{9}{n}\right)^n$. Apply the Root Test, since everything is raised to the nth power:

$$\lim_{n\to\infty}\sqrt[n]{\left(\frac{9}{n}\right)^n}=\lim_{n\to\infty}\frac{9}{n}=0$$

Since $0<1$, this series converges.

6. Divergent. It is clearly an alternating series whose terms grow smaller as n gets larger and larger. Even though the numerator of the series gets bigger by 1 for each consecutive term, the denominator grows by more than a factor of 3. However, the series fails the nth term divergence test since:

$$\lim_{n\to\infty}\frac{n}{3n+1}=\frac{1}{3}$$

The limit at infinity *must* equal 0 for the series to converge, and it just doesn't.

7. Convergent. The Alternating Series Test fails, since the limit at infinity results in an indeterminate answer. Thus, we'll test for absolute convergence and check out the series $\displaystyle\sum_{n=0}^{\infty}\frac{4\cdot2^n}{3^n}$. If you rewrite the series as $\displaystyle\sum_{n=0}^{\infty}4\cdot\left(\frac{2}{3}\right)^n$, it's obvious that you're dealing with a geometric series whose ratio is $\frac{2}{3}$. Since $\left|\frac{2}{3}\right|<1$, the series $\displaystyle\sum_{n=0}^{\infty}4\cdot\left(\frac{2}{3}\right)^n$ converges, which means that $\displaystyle\sum_{n=0}^{\infty}\frac{(-1)^n\cdot4\cdot2^n}{3^n}$ converges automatically.

Chapter 24

1. ∞. Begin with the Ratio Test for absolute convergence:

$$\lim_{n \to \infty} \left| \frac{5^{n+1} x^{n+1}}{(n+1)!} \cdot \frac{n!}{5^n x^n} \right|$$

$$\lim_{n \to \infty} \left| \frac{5x}{n+1} \right|$$

As n approaches infinity, $\left| \frac{5}{n+1} \right|$ approaches 0. Remember that the limit has to be less than 1 in order for the series to converge:

$$|0 \cdot x| < 1$$

Hold the phone! No matter what x is, you get 0 on the left side of the inequality, and 0 is *always* less than 1. Therefore, this series will converge regardless of the value of x, meaning that the series converges on the interval $(-\infty, \infty)$.

2. $(1,3]$. First, find the radius of convergence using the Ratio Test for this power series centered around $x = 2$. The $(-1)^{n+1}$ term will vanish because of the absolute value signs:

$$\lim_{n \to \infty} \left| \frac{(x-2)^{n+2}}{n+1} \cdot \frac{n}{(x-2)^{n+1}} \right|$$

$$= \lim_{n \to \infty} \left| \frac{n(x-2)}{n+1} \right| = \left| 1 \cdot (x-2) \right|$$

Remember, the series only converges when $|x-2| < 1$. Because this is in the form $|x - c| < r$, you know that the radius of convergence is 1, and the series converges inside the interval $(2-1, 2+1) = (1,3)$. Now, we should check to see if the series converges at the endpoints. Start with $x = 1$:

$$\sum_{n=1}^{\infty} \frac{(-1)^{n+1}(-1)^{n+1}}{n} = \sum_{n=1}^{\infty} \frac{(-1 \cdot -1)^{n+1}}{n} = \sum_{n=1}^{\infty} \frac{1^{n+1}}{n}$$

This is the harmonic series, which is divergent. Now, check the other endpoint, $x = 3$:

$$\sum_{n=1}^{\infty} \frac{(-1)^{n+1}(1)^{n+1}}{n} = \sum_{n=1}^{\infty} \frac{(-1 \cdot 1)^{n+1}}{n} = \sum_{n=1}^{\infty} \frac{(-1)^{n+1}}{n}$$

This series converges according to the Alternating Series Test. Therefore, we must include $x = 3$ in the interval of convergence, which will be $(1,3]$.

3. 1.284017. No matter what derivative you take of $f(x) = e^x$, $f^{(n)}(0) = e^0 = 1$. Therefore, the Maclaurin polynomial will be ...

$$e^x \approx \frac{1 \cdot x^0}{0!} + \frac{1 \cdot x^1}{1!} + \frac{1 \cdot x^2}{2!} + \frac{1 \cdot x^3}{3!} + \frac{1 \cdot x^4}{4!}$$

$$\approx 1 + x + \frac{x^2}{2} + \frac{x^3}{6} + \frac{x^4}{24}$$

Simply plug $x = 0.25$ into the polynomial to get your approximation:

$$e^{0.25} \approx 1 + (0.25) + \frac{(0.25)^2}{2} + \frac{(0.25)^3}{6} + \frac{(0.25)^4}{24}$$

$$\approx 1.284017$$

If you're curious, $e^{0.25}$ is actually equal to 1.284025.

4. 2.049375. You need to find up to the second derivative of $f(x) = x^{1/2}$ evaluated at 4:

$$f(x) = \sqrt{x} \qquad f(4) = 2$$

$$f'(x) = \frac{1}{2\sqrt{x}} \qquad f'(4) = \frac{1}{4}$$

$$f''(x) = -\frac{1}{4x^{3/2}} \qquad f''(4) = -\frac{1}{32}$$

Plug these values into the Taylor series expansion centered at $c = 4$:

$$f(4) + \frac{f'(4)(x-4)}{1!} + \frac{f''(4)(x-4)^2}{2!}$$

$$= 2 + \frac{1}{4} \cdot \frac{(x-4)}{1} + \left(-\frac{1}{32}\right)\frac{(x-4)^2}{2}$$

$$= 2 + \frac{x-4}{4} - \frac{(x-4)^2}{64}$$

Plug $x = 4.2$ into this polynomial to get an approximation for $\sqrt{4.2}$:

$$\sqrt{4.2} \approx 2 + \frac{4.2 - 4}{4} - \frac{\left(4.2 - 4\right)^2}{64}$$

$$\approx 2 + \frac{0.2}{4} - \frac{\left(0.2\right)^2}{64} = 2.049375$$

If you're wondering, the actual value of $\sqrt{4.2}$ is 2.049390.

Appendix B

Glossary

absolute convergence   Describes a series $\sum a_n$ if the corresponding series $\sum |a_n|$ converges; it is a method of determining whether or not a series containing negative terms converges if you cannot use the Alternating Series Test.

absolute extreme point   The highest or lowest point on a graph.

acceleration   The rate of change of velocity.

accumulation function   A function defined by a definite integral; it has a variable in one or both of its limits of integration.

alternating series   Series whose consecutive terms alternate between positive and negative.

Alternating Series Test   If $\sum a_n$ is an alternating series, and ...

  1) Every term of the series is less than or equal to the term preceding it; and

  2) $\lim\limits_{n \to \infty} a_n = 0$;

... then $\sum a_n$ converges.

antiderivative   The opposite of the derivative; if $f(x)$ is an antiderivative of $g(x)$, then $\int g(x)dx = f(x)$.

antidifferentiation   The process of creating an antiderivative or integral.

asymptote   A line representing an unattainable value that shapes a graph; because the graph cannot achieve the value, the graph typically bends toward that line forever and ever.

average value of a function   The value, $f(c)$, guaranteed by the Mean Value Theorem for Integration found via the equation $f(c) = \dfrac{\int_a^b f(x)dx}{b-a}$.

Chain Rule   The derivative of the composite function $h(x) = f(g(x))$ is $h'(x) = f'(g(x)) \cdot g'(x)$.

cofunction   Trigonometric functions with the same name, apart from the prefix "co-," like sine and cosine or tangent and cotangent.

Comparison Test   Given two infinite series with positive terms $\sum a_n$ and $\sum b_n$, so that every term of $\sum a_n$ is less than or equal to the corresponding term in $\sum b_n$:

1) If $\sum b_n$ converges, then $\sum a_n$ converges.

2) If $\sum a_n$ diverges, then $\sum b_n$ diverges.

concavity   Describes how a curve bends; a curve that can hold water poured into it from the top of the graph is said to be concave up, whereas one that cannot hold water is said to be concave down.

conjugate   A binomial whose middle sign is the opposite of another binomial with the same terms (e.g., $3 + \sqrt{x}$ and $3 - \sqrt{x}$ are conjugates).

constant   A polynomial of degree 0; a real number.

constant of integration   The unknown constant which results from an indefinite integral, usually written as C in your solution; it is a required piece of all indefinite integral solutions.

continuous   A function $f(x)$ is continuous at $x = c$ if $\lim\limits_{x \to c} f(x) = f(c)$.

convergent sequence   Has elements that approach, but never reach, some limiting value; if $\lim\limits_{n \to \infty} a_n$ exists (i.e., is some real number), then the sequence $\{a_n\}$ converges.

coterminal angles   Angles which have the same function value, because the space between them is a multiple of the function's period.

critical number   An x-value where a graph *could* change direction, indicated mathematically by a zero derivative or a point of nondifferentiability.

critical point   A coordinate pair where a graph *could* change direction, indicated mathematically by a zero derivative or a point of nondifferentiability.

cubic   A polynomial of degree three.

definite integral   An integral which contains limits of integration; its solution is a real number.

degree   The largest exponent in a polynomial.

derivative   The derivative of a function $f(x)$ at $x = c$ is the slope of the tangent line to f at $x = c$, usually written $f'(c)$.

difference quotient   One of two formulas which defines a derivative:

$$f'(x) = \lim_{\Delta x \to \infty} \frac{f(x + \Delta x) - f(x)}{\Delta x} \ \text{ or } \ f'(c) = \lim_{x \to c} \frac{f(x) - f(c)}{x - c}.$$

differentiable   Possessing a derivative at the specific x-value; if a function does not have a derivative at the given point, it is said to be "nondifferentiable" there.

differential equation   An equation containing a derivative.

disk method   The volume of a solid generated by rotating an area about the x-axis in three dimensions is equal to $\pi \int_a^b \left(r(x) \right)^2 dx$, where $r(x)$ is the radius of rotation and a and b are the x-values bounding the area to be rotated.

displacement   The total change in position counting only the beginning and ending position; if the object in question changes direction any time during that interval of time, it does not correctly reflect the total distance traveled.

divergent   A sequence or series which does not converge (i.e., is not bounded and therefore has no limiting value).

domain   The set of possible inputs for a function.

essential discontinuity   See *infinite discontinuity*.

Euler's Method   A technique used to approximate values on the solution graph to a differential equation when you can't actually find the specific solution to the differential equation via separation of variables.

everywhere continuous   A function which is continuous at each x in its domain.

exponential growth and decay   A population grows or decays exponentially if its rate of change is proportional to the population itself; i.e., $\frac{dP}{dt} = k \cdot P$, where k is a constant and P is the size of the population.

extrema point   The plural form of "extreme points."

extreme point   A high or low point in the curve, a *maximum* or a *minimum*, respectively; it represents an extreme value of the graph, whether extremely high or extremely low, in relation to the points surrounding it.

Extreme Value Theorem   If a function $f(x)$ is continuous on the closed interval $[a,b]$, then f has an absolute maximum and an absolute minimum on $[a,b]$.

factorial   Number with an exclamation point beside it, like 4!; it equals the product of the number and all of the integers preceding it down to and including 1: $4! = 4 \cdot 3 \cdot 2 \cdot 1$.

factoring   Reversing the process of multiplication. The results of the factoring process can be multiplied together to get the original quantity.

family of solutions   Any mathematical solution containing "+C"; it compactly represents an infinite number of possible solutions, each differing only by a constant.

function   A relation such that every input has only one matching output.

geometric series   A series that has the form $\sum\limits_{n=0}^{\infty} ar^n$, where a and r are constants; it converges if $0 < |r| < 1$.

greatest common factor   The largest quantity by which all the terms of an expression can be divided evenly.

harmonic series   The divergent p-series with $p = 1$, i.e., $\sum\limits_{n=1}^{\infty} \frac{1}{n}$.

implicit differentiation   Allows you to find the slope of a tangent line when the equation in question cannot be solved for y.

indefinite integral   An integral which does not contain limits of integration; its solution is the antiderivative of the expression (and must contain a constant of integration).

indeterminate form   An expression whose value is unclear; the most common indeterminate forms are $\pm\frac{\infty}{\infty}$, $\frac{0}{0}$, and $0 \cdot \infty$.

infinite discontinuity   Discontinuity caused by a vertical asymptote (also called "essential discontinuity").

inflection points   Points on a graph where its concavity changes.

inner radius   Radius of rotation used in the washer method that extends from the rotational axis to the inner edge of the region.

integer   A number without a decimal or fractional part.

integral   The opposite of the derivative; if $f(x)$ is the integral of $g(x)$, then $\int g(x)dx = f(x)$.

Integral Test   The positive series $\sum\limits_{n=1}^{\infty} a_n$ converges if the improper integral $\int_1^{\infty} a_n \, dn$ has a finite value; if the integral diverges (increases without bound), so does the series.

integration   The process of creating an antiderivative or integral.

integration by parts   Allows you to rewrite the integral $\int u \, dv$ (where u is an easily differentiated function and dv is one easily integrated) as $uv - \int v \, du$.

intercept   Numeric value where a graph hits either the x- or y-axis.

Intermediate Value Theorem   If a function $f(x)$ is continuous on the closed interval $[a,b]$, then for every real number d between $f(a)$ and $f(b)$, there exists a c between a and b so that $f(c) = d$.

interval of convergence   The interval on which a power series converges; it is found by first determining the radius of convergence r (so that the series converges for all numbers between $c - r$ and $c + r$) and then checking for convergence at the endpoints $c - r$ and $c + r$.

irrational root   An x-intercept that cannot be written as a fraction.

jump discontinuity   Occurs when no general limit exists at the given x-value.

left-hand limit   The height a function intends to reach as you approach the given x-value *from* the left.

left sum   A Riemann approximation where the heights of the rectangles are defined by the value of the function at the left-hand side of each interval.

L'Hôpital's Rule   If a limit results in an indeterminate form after substitution, you can take the derivatives of the numerator and denominator of the fraction separately without changing the limit's value (i.e., $\lim\limits_{x \to c} \dfrac{f(x)}{g(x)} = \lim\limits_{x \to c} \dfrac{f'(x)}{g'(x)}$).

limit   The height a function *intends* to reach at a given x-value, whether or not it actually reaches it.

Limit Comparison Test   Given the positive infinite series $\sum a_n$ and $\sum b_n$, if $\lim\limits_{n \to \infty} \dfrac{a_n}{b_n} = N$, where N is a positive and finite number, then $\sum a_n$ and $\sum b_n$ either both converge or both diverge.

limits of integration   Small numbers next to the integral sign, indicating the boundaries when calculating area under the curve; in the expression $\int_1^3 2x\, dx$, the limits of integration are 1 and 3.

linear approximation   The equation of a tangent line to a function used to help approximate the function's values lying close to the point of tangency.

linear expression   A polynomial of degree 1.

logistic growth   Begins quickly (it in fact looks like exponential growth) but eventually slows and levels off to some limiting value; most natural phenomena, including population and sales graphs, follow this pattern rather than exponential growth.

Maclaurin series   The series $\sum\limits_{n=0}^{\infty} \dfrac{f^{(n)}(0)x^n}{n!}$, which gives a good approximation for the function $f(x)$'s values near $x = 0$; you typically only use a finite number of terms, which results in a polynomial, rather than an infinite series.

Mean Value Theorem   If a function $f(x)$ is continuous and differentiable on a closed interval $[a,b]$, then there exists a point c, $a \le c \le b$, so that $f'(c) = \dfrac{f(b) - f(a)}{b - a}$.

Mean Value Theorem for Integration   If a function $f(x)$ is continuous on the interval $[a,b]$, then there exists a c, $a \le c \le b$, such that $(b - a) \cdot f(c) = \int_a^b f(x)dx$.

midpoint sum   A Riemann approximation where the heights of the rectangles are defined by the value of the function at the midpoint of each interval.

nondifferentiable   Not possessing a derivative at the given value.

nonremovable discontinuity   A point of discontinuity for which no limit exists (e.g., infinite or jump discontinuity).

normal line   The line perpendicular to a function's tangent line at the point of tangency.

nth term divergence test   The infinite series $\sum a_n$ is divergent if $\lim\limits_{n \to \infty} a_n \neq 0$.

optimizing   Finding the maximum or minimum value of a function given a set of circumstances.

outer radius   Radius of rotation used in the washer method that extends from the rotational axis to the outer edge of the region.

p-series   Has the form $\sum\limits_{n=1}^{\infty} \frac{1}{n^p}$, where p is a constant; it converges if $p > 1$, but diverges for all other values of p.

parameter   A variable into which you plug numeric values to find coordinates on a parametric equation graph.

parametric equations   Pairs of equations, usually in the form of "$x =$" and "$y =$," that define points of a graph in terms of yet another variable, usually t.

partial fraction decomposition   A method of rewriting a fraction as a sum and difference of smaller fractions, whose denominators are factors of the original, larger denominator.

period   The amount of horizontal space it takes until a periodic function repeats itself.

periodic function   A function whose values repeat over and over, at the same rate and at the same intervals in time.

point discontinuity   Occurs where a general limit exists, but the function value is not defined.

point-slope form   A line containing the point (x_1, y_1) and having slope m has equation $y - y_1 = m(x - x_1)$.

position equation   A mathematical model which outputs an object's position at a given time, t.

positive series   A series that contains only positive terms.

Power Rule   The derivative of the expression ax^n, where a and n are real numbers, is $(an)x^{n-1}$.

Power Rule for Integration   The integral of a single variable to some power is found by adding 1 to the existing exponent and dividing the entire expression by the new exponent: $\int x^n\,dx = \frac{x^{n+1}}{n+1} + C$ (provided $n \neq -1$).

power series   A power series centered at $x = c$ has form $\sum\limits_{n=0}^{\infty} a_n(x-c)^n$ and is used to approximate function values close to $x = c$.

Product Rule   The derivative of $f(x)g(x)$, where f and g are variable expressions, is $f(x) \cdot g'(x) + f'(x) \cdot g(x)$.

quadratic   A polynomial of degree two.

Quotient Rule   If $h(x) = \frac{f(x)}{g(x)}$, where $f(x)$ and $g(x)$ are two differentiable functions, then

$$h'(x) = \frac{g(x) \cdot f'(x) - f(x) \cdot g'(x)}{\left(g(x)\right)^2}.$$

radius of convergence   If a power series centered at c has radius of convergence r, then that series will converge for all x-values on the interval $|x - c| < r$; in other words, all x's falling in the interval $(c - r, c + r)$ cause the power series to converge.

radius of rotation   A line segment extending from the axis of rotation to the edge of the area being rotated.

range   The set of possible outputs for a function.

ratio   The r term in the geometric series $\sum\limits_{n=0}^{\infty} ar^n$.

Ratio Test   If $\sum a_n$ is an infinite series of positive terms, and $\lim\limits_{n \to \infty} \frac{a_{n+1}}{a_n} = L$ (where L is a real number), then:

1) $\sum a_n$ converges if $L < 1$,
2) $\sum a_n$ diverges if $L > 1$ or if $L = \infty$, and
3) If $L = 1$, no conclusion can be drawn from the Ratio Test.

reciprocal   The fraction flipped upside down (e.g., the reciprocal of $\frac{7}{4}$ is $\frac{4}{7}$).

related rates   A problem that uses a known variable's rate of change to compute the rate of change for another variable in the problem.

relation   A collection of related numbers, most often described by an equation.

relative extreme point   Occurs when that point is higher or lower than all of the points in the immediate surrounding area; visually, a relative maximum is the peak of a hill in the graph, and a relative minimum is the lowest point of a dip in the graph.

removable discontinuity   A point of discontinuity for which a limit exists (i.e., point discontinuity).

repeating factor   A factor in the denominator of a fraction raised to a power in the process of partial fraction decomposition.

representative radius   Extends from one edge of a region to the opposite edge; used in the shell method.

Riemann sum   An approximation for the area beneath a curve that is achieved by adding the areas of rectangles.

right-hand limit   The function's intended height as you approach the given x-value *from* the right.

right sum   A Riemann approximation where the heights of the rectangles are defined by the value of the function at the right-hand side of each interval.

Rolle's Theorem   If a function $f(x)$ is continuous and differentiable on a closed interval $[a,b]$ and $f(a) = f(b)$, then there exists a c between a and b so that $f'(c) = 0$.

Root Test   If $\sum a_n$ is an infinite series of positive terms, and $\lim_{n \to \infty} \sqrt[n]{a_n} = L$, then:

1) $\sum a_n$ converges if $L < 1$,

2) $\sum a_n$ diverges if $L > 1$ or if $L = \infty$, and

3) If $L = 1$, no conclusion can be drawn from the Root Test (just like the Ratio Test).

secant line   A line which cuts through a graph, usually intersecting it in multiple locations.

separation of variables   A technique used to solve basic differential equations; in it, you move the different variables of the equation to different sides of the equal sign in order to integrate each side of the equation separately and so reach a solution.

sequence   A list of numbers generated by some mathematical rule typically expressed in terms of n; in order to construct the sequence, you plug in consecutive integer values into n.

series   The sum of the terms of a sequence; the series indicates which terms are to be added via its sigma notation boundaries.

shell method   A procedure used to calculate the volume of a rotational solid, whether it's completely solid or partially hollow; it is the only rotational volume calculation method that uses radii parallel to, rather than perpendicular to, the axis of rotation.

sign graph   See *wiggle graph*.

Simpson's Rule   The approximate area under the curve $f(x)$ on the closed interval $[a,b]$ using an even number of subintervals, n, is ...

$$\frac{b-a}{3n}\left(f(a)+4f(x_1)+2f(x_2)+\cdots+2f(x_{n-2})+4f(x_{n-1})+f(b)\right)$$

slope   Numeric value that (via its sign and how large it is) describes the overall "slanti-ness" of a line.

slope field   A tool to help visualize the solution to a differential equation; it is made up of a collection of line segments centered at points whose slopes are the values of the dif-ferential equation evaluated at those points.

slope-intercept form   A line with slope m and y-intercept b has equation $y = mx + b$.

speed   The absolute value of velocity.

symmetric function   Looks like a mirror image of itself, typically across the x-axis, y-axis, or about the origin.

tangent line   A line which skims across the curve, hitting it once in the indicated loca-tion.

Taylor series   Series that have the form $\sum\limits_{n=0}^{\infty}\frac{f^{(n)}(c)(x-c)^n}{n!}$ and give good estimations for values of $f(x)$ near $x = c$.

telescoping series   Series which contain an infinite number of terms and their oppo-sites, resulting in almost all of the terms in the series canceling out.

Trapezoidal Rule   The approximate area beneath a curve $f(x)$ on the interval $[a,b]$ using n trapezoids is ...

$$\frac{b-a}{2n}\left(f(a)+2f(x_1)+2f(x_2)+2f(x_3)\cdots+2f(x_{n-1})+f(b)\right)$$

u-substitution   Integration technique that helps you integrate expressions containing both a function and its derivative.

velocity   The rate of change of position; it includes a component of direction, and there-fore, may be negative.

vertical line test   Tells you whether or not a graph is a function; if any vertical line can be drawn through the graph which intersects that graph more than once, then the graph in question cannot be a function.

washer method   A procedure used to calculate the volume of a rotational solid even if part of it is hollow.

wiggle graph   A segmented number line that describes the direction of a function.

Index

A Little Knowledge Goes a Long Way ...

Check Out These Best-Selling COMPLETE IDIOT'S GUIDES

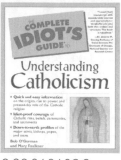

0-02-863639-2
$16.95

0-02-862743-1
$16.95

0-02-862728-8
$16.95

0-02-864339-9
$18.95

0-02-864244-9
$21.95 w/CD-ROM

0-02-862415-7
$18.95

0-02-864316-X
$24.95 w/CD-ROM

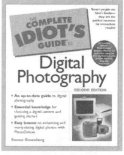

0-02-864235-X
$24.95 w/CD-ROM

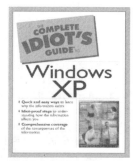

0-02-864232-5
$19.95

More than *400 titles* in *26 different categories*
Available at booksellers everywhere

ALPHA